Instructor's Solutions Manual

for

Probability and Statistics for Engineers and Scientists

Anthony J. Hayter

Georgia Institute of Technology

PWS Publishing Company

An International Thomson Publishing Company

Boston • Albany • Bonn • Cincinnati • Detroit • London •Madrid • Melbourne • Mexico City
New York • Paris • San Francisco • Singapore • Tokyo • Toronto • Washington

PWS PUBLISHING COMPANY
20 Park Plaza, Boston, Massachusetts 02116-4324

International Thomson Publishing
The trademark ITP is used under license

For more information, contact:

PWS Publishing Co.
20 Park Plaza
Boston, MA 02116

International Thomson Publishing Europe
Berkshire House I68-I73
High Holborn
London WC1V 7AA
England

Thomas Nelson Australia
102 Dodds Street
South Melbourne, 3205
Victoria, Australia

Nelson Canada
1120 Birchmont Road
Scarborough, Ontario
Canada M1K 5G4

International Thomson Editores
Campos Eliseos 385, Piso 7
Col. Polanco
11560 Mexico D.F., Mexico

International Thomson Publishing GmbH
Konigswinterer Strasse 418
53227 Bonn, Germany

International Thomson Publishing Asia
221 Henderson Road
#05-10 Henderson Building
Singapore 0315

International Thomson Publishing Japan
Hirakawacho Kyowa Building, 31
2-2-1 Hirakawacho
Chiyoda Ku, Tokyo 102
Japan

ISBN 0-534-95612-2

Text and Cover Printer: Financial Publishing

Printed and bound in the United States of America
96 97 98 99---10 9 8 7 6 5 4 3 2 1

Contents

Instructor Solution Manual

This instructor solution manual to accompany "Probability and Statistics for Engineers and Scientists" by Anthony Hayter provides worked solutions and answers to *all* of the problems given in the textbook. The student solution manual provides worked solutions and answers to only the odd-numbered problems given at the end of the chapter sections. In addition to the material contained in the student solution manual, this instructor manual therefore provides worked solutions and answers to the even-numbered problems given at the end of the chapter sections together with all of the supplementary problems at the end of each chapter.

You are reminded that the data diskette provided with the textbook contains all of the data sets used in the problems. The data sets are provided in ASCII format and as Minitab worksheets and are referenced by the figure number where they appear in the textbook.

Chapter 1

Probability Theory

1.1 Probabilities

1.1.1 $\mathcal{S} = \{$(head, head, head), (head, head, tail), (head, tail, head), (head, tail, tail), (tail, head, head), (tail, head, tail), (tail, tail, head), (tail, tail, tail)$\}$

1.1.2 $\mathcal{S} = \{0$ females, 1 female, 2 females, 3 females, $\ldots, n$ females$\}$

1.1.3 $\mathcal{S} = \{$all U. S. citizens born in the U.S.A. of at least 35 years of age who have not already served two terms as president $\}$

1.1.4 $\mathcal{S} = \{$Jan. 1, Jan. 2, Jan. 3, ... , Feb. 29, ... , Dec. 31$\}$

1.1.5 $\mathcal{S} = \{$(on time, satisfactory), (on time, unsatisfactory), (late, satisfactory), (late, unsatisfactory)$\}$

1.1.6 $\mathcal{S} = \{$(red, shiny), (red, dull), (blue, shiny), (blue, dull)$\}$

1.1.7 (a) $\frac{p}{1-p} = 1 \quad \Rightarrow \quad p = 0.5$

   (b) $\frac{p}{1-p} = 2 \quad \Rightarrow \quad p = \frac{2}{3}$

   (c) $p = 0.25 \quad \Rightarrow \quad \frac{p}{1-p} = \frac{1}{3}$

1.1.8 $0.13 + 0.24 + 0.07 + 0.38 + P(V) = 1 \quad \Rightarrow \quad P(V) = 0.18.$

1.1.9 $0.08 + 0.20 + 0.33 + P(IV) + P(V) = 1 \quad \Rightarrow \quad P(IV) + P(V) = 1 - 0.61 = 0.39.$
Thus, $0 \leq P(V) \leq 0.39.$
If $P(IV) = P(V)$, then $P(V) = 0.195.$

7

1.1.10 $P(I) = 2 \times P(II)$ and $P(II) = 3 \times P(III) \Rightarrow P(I) = 6 \times P(III)$.

Then $P(I) + P(II) + P(III) = 1 \Rightarrow (6 \times P(III)) + (3 \times P(III)) + P(III) = 1$.

Hence $P(III) = \frac{1}{10}$, $P(II) = 3 \times P(III) = \frac{3}{10}$ and $P(I) = 6 \times P(III) = \frac{6}{10}$.

1.2   Events

1.2.1   (a) $0.13 + P(b) + 0.48 + 0.02 + 0.22 = 1 \quad \Rightarrow \quad P(b) = 0.15.$

(b) A = {c, d} so $P(A) = P(c) + P(d) = 0.48 + 0.02 = 0.50.$

(c) $P(A') = 1 - P(A) = 1 - 0.5 = 0.50.$

1.2.2   (a) $P(A) = P(b) + P(c) + P(e) = 0.27$ so $P(b) + 0.11 + 0.06 = 0.27$ and hence $P(b) = 0.10.$

(b) $P(A') = 1 - P(A) = 1 - 0.27 = 0.73.$

(c) $P(A') = P(a) + P(d) + P(f) = 0.73$ so $0.09 + P(d) + 0.29 = 0.73$ and hence $P(d) = 0.35.$

1.2.3   Over a four year period including one leap year, the number of days is $(3 \times 365) + 366 = 1,461$. The number of January days is $4 \times 31 = 124$ and the number of February days is $(3 \times 28) + 29 = 113$. The answers are therefore $\frac{124}{1,461}$ and $\frac{113}{1,461}$.

1.2.4 $S = \{1, 2, 3, 4, 5, 6\}$ and Prime = $\{1, 2, 3, 5\}$.
All the events in S are equally likely to occur and each has a probability of $\frac{1}{6}$.
Thus $P(\text{Prime}) = P(1) + P(2) + P(3) + P(5) = \frac{4}{6} = \frac{2}{3}$

1.2.5   See Figure 1.10. The event that the score on at least one of the two dice is a prime number consists of the following 32 outcomes:

{(1,1), (1,2), (1,3), (1,4), (1,5), (1,6), (2,1), (2,2), (2,3), (2,4) (2,5), (2,6), (3,1), (3,2), (3,3), (3,4), (3,5), (3,6), (4,1), (4,2), (4,3), (4,5), (5,1), (5,2), (5,3), (5,4), (5,5), (5,6), (6,1), (6,2), (6,3), (6,5)}.

Since each outcome in S is equally likely to occur with a probability of $\frac{1}{36}$,
P(at least one score is a prime number) $= 32 \times \frac{1}{36} = \frac{32}{36} = \frac{8}{9}.$

The complement of this event is the event that neither score is a prime number which includes the following four outcomes:

{(4,4), (4,6), (6,4), (6,6)}.

P(neither score prime) $= \frac{1}{36} + \frac{1}{36} + \frac{1}{36} + \frac{1}{36} = \frac{1}{9}.$

1.2.6   In Figure 1.10, let (x, y) represent the outcome that the score on the red die is x and the score on the blue die is y. The event that the score on the red die is *strictly greater* than the score on the blue die consists of the following 15 outcomes:

{(2,1), (3,1), (3,2), (4,1), (4,2), (4,3), (5,1), (5,2), (5,3), (5,4), (6,1), (6,2), (6,3),

(6,4), (6,5)}

The probability of each outcome is $\frac{1}{36}$, so the required probability is $15 \times \frac{1}{36} = \frac{5}{12}$. This probability is less than 0.5 because of the possibility that both scores are equal. The complement of this event is the event that the red die has a score *less than or equal* to the score on the blue die with a probability of $1 - \frac{5}{12} = \frac{7}{12}$.

1.2.7 $P(\spadesuit \text{ or } \clubsuit) = P(A\spadesuit) + P(K\spadesuit) + \ldots + P(2\spadesuit) + P(A\clubsuit) + P(K\clubsuit) + \ldots + P(2\clubsuit)$
$= \frac{1}{52} + \ldots + \frac{1}{52} = \frac{26}{52} = \frac{1}{2}$.

1.2.8 $P(\text{draw an ace}) = P(A\spadesuit) + P(A\clubsuit) + P(A\Diamond) + P(A\heartsuit) = \frac{1}{52} + \frac{1}{52} + \frac{1}{52} + \frac{1}{52}$
$= \frac{4}{52} = \frac{1}{13}$.

1.2.9   (a) Let the four players be named A, B, C, and T for Terica. The notation (X, Y) indicates that player X is the winner and player Y is the runner up. The sample space consists of the 12 outcomes

$S = \{(A,B), (A,C), (A,T), (B,A), (B,C), (B,T), (C,A), (C,B), (C,T), (T,A), (T,B), (T,C)\}$

The event *Terica is winner* consists of the 3 outcomes, $\{(T,A), (T,B), (T,C)\}$. Since each outcome in S is equally likely to occur with a probability of $\frac{1}{12}$, P(Terica is winner) $= \frac{3}{12} = \frac{1}{4}$.

(b) The event *Terica is winner or runner up* consists of 6 out of the 12 outcomes. Thus, P(Terica is winner or runner up) $= \frac{6}{12} = \frac{1}{2}$.

1.2.10   (a) See Figure 1.24.
P(Type I battery lasts longest) = P((II, III, I)) + P((III, II, I))
= 0.39 + 0.03 = 0.42.

(b) P(Type I battery lasts shortest) = P((I, II, III)) + P((I, III, II))
= 0.11 + 0.07 = 0.18.

(c) P(Type I battery does not last longest) = 1 − P(Type I battery lasts longest)
= 1 − 0.42 = 0.58.

(d) P(Type I battery last longer than Type II)
= P((II, I, III)) + P((II, III, I)) + P((III, II, I))
= 0.24 + 0.39 + 0.03 = 0.66.

1.2.11   (a) See Figure 1.25.
The event *both assembly lines are shut down* consists of the single outcome $\{(S,S)\}$.
Hence P(both assembly lines are shut down) = 0.02.

(b) The event *neither assembly line is shut down* consists of the outcomes
{(P,P), (P,F), (F,P), (F,F)}.
Hence P(neither assembly line is shut down)
= P((P,P)) + P((P,F)) + P((F,P)) + P((F,F))
= 0.14 + 0.2 + 0.21 + 0.19 = 0.74.

(c) The event *at least one assembly line is at full capacity* consists of the outcomes
{(S,F), (P,F), (F,F), (F,S), (F,P)}.
Hence P(at least one assembly line is at full capacity)
= P((S,F)) + P((P,F)) + P((F,F)) + P((F,S)) + P((F,P))
= 0.05 + 0.2 + 0.19 + 0.06 + 0.21 = 0.71.

(d) The event *exactly one assembly line at full capacity* consists of the outcomes
{(S,F), (P,F), (F,S), (F,P)}.
Hence P(exactly one assembly line at full capacity)
= P((S,F)) + P((P,F)) + P((F,S)) + P((F,P))
= 0.05 + 0.20 + 0.06 + 0.21 = 0.52.

The complement of *neither assembly line is shut down* is the event *at least one assembly line is shut down* which consists of the outcomes
{(S,S), (S,P), (S,F), (P,S), (F,S)}.

The complement of *at least one assembly line is at full capacity* is the event *neither assembly line is at full capacity* which consists of the outcomes
{(S,S), (S,P), (P,S), (P,P)}.

1.2.12 The sample space is
S = {(H,H,H), (H,T,H), (H,T,T), (H,H,T), (T,H,H), (T,H,T), (T,T,H), (T,T,T)}
with each outcome being equally likely with a probability of $\frac{1}{8}$.

The event *two heads obtained in succession* consists of the three outcomes
{(H,H,H), (H,H,T), (T,H,H)}
so that P(two heads in succession) = $\frac{3}{8}$.

1.3 Combinations of Events

1.3.1 The event A contains the outcome 0 while the empty set does not contain any outcomes.

1.3.2 (a) See Figure 1.55.
$P(B) = 0.01 + 0.02 + 0.05 + 0.11 + 0.08 + 0.06 + 0.13 = 0.46.$

(b) $P(B \cap C) = 0.02 + 0.05 + 0.11 = 0.18.$

(c) $P(A \cup C) = 0.07 + 0.05 + 0.01 + 0.02 + 0.05 + 0.08 + 0.04 + 0.11 + 0.07 + 0.11 = 0.61.$

(d) $P(A \cap B \cap C) = 0.02 + 0.05 = 0.07.$

(e) $P(A \cup B \cup C) = 1 - 0.03 - 0.04 - 0.05 = 0.88.$

(f) $P(A' \cap B) = 0.08 + 0.06 + 0.11 + 0.13 = 0.38.$

(g) $P(B' \cup C) = 0.04 + 0.03 + 0.05 + 0.11 + 0.05 + 0.02 + 0.08 + 0.04 + 0.11 + 0.07 + 0.07 + 0.05 = 0.72.$

(h) $P(A \cup (B \cap C)) = 0.07 + 0.05 + 0.01 + 0.02 + 0.05 + 0.08 + 0.04 + 0.11 = 0.43.$

(i) $P((A \cup B) \cap C) = 0.11 + 0.05 + 0.02 + 0.08 + 0.04 = 0.30.$

(j) $P(A' \cup C)' = 1 - P(A' \cup C)$
$P(A' \cup C) = 0.04 + 0.03 + 0.05 + 0.08 + 0.06 + 0.13 + 0.11 + 0.11 + 0.07 + 0.02 + 0.05 + 0.08 + 0.04 = 0.87.$
$P(A' \cup C)' = 1 - 0.87 = 0.13.$

1.3.4 (a) $A \cap B$ = {females with black hair}.

(b) $A \cup C'$ = {all females together with any man who does not have brown eyes}.

(c) $A' \cap B \cap C$ = {males with black hair and brown eyes}.

(d) $A \cap (B \cup C)$ = {females with either black hair or brown eyes or both}.

1.3.5 Yes, a card must be drawn from either a red suit or a black suit, but it cannot be from both at the same time.
No, the ace of hearts could have been drawn.

1.3.6 $P(A \cup B) = P(A) + P(B) - P(A \cap B) \leq 1$ so $P(B) \leq 1 - 0.4 + 0.3 = 0.9.$
Also, $P(B) \geq P(A \cap B) = 0.3$, so $0.3 \leq P(B) \leq 0.9.$

1.3.7 Since $P(A \cup B) = P(A) + P(B) - P(A \cap B)$, it follows that
$P(B) = P(A \cup B) - P(A) + P(A \cap B) = 0.8 - 0.5 + 0.1 = 0.4.$

1.3.8 $S = \{1, 2, 3, 4, 5, 6\}$ where each outcome is equally likely with a probability of $\frac{1}{6}$.
The events A, B, and B' are A = $\{2, 4, 6\}$, B = $\{1, 2, 3, 5\}$ and $B' = \{4, 6\}$.

    (a) $A \cap B = \{2\}$ so $P(A \cap B) = \frac{1}{6}$.

    (b) $A \cup B = \{1, 2, 3, 4, 5, 6\}$ so $P(A \cup B) = 1$.

    (c) $A \cap B' = \{4, 6\}$ so $P(A \cap B') = \frac{2}{6} = \frac{1}{3}$.

1.3.9   Yes, the three events are mutually exclusive because the selected card can only be from one suit.

Therefore $P(A \cup B \cup C) = P(A) + P(B) + P(C) = \frac{1}{4} + \frac{1}{4} + \frac{1}{4} = \frac{3}{4}$.

A' is the event *a heart is not obtained* (or similarly the event *a club, spade, or diamond is obtained*). Hence B is just a subset of A'.

1.3.10   (a) $A \cap B = \{A\heartsuit, A\diamondsuit\}$

    (b) $A \cup C = \{A\heartsuit, A\diamondsuit, A\clubsuit, A\spadesuit, K\heartsuit, K\diamondsuit, K\clubsuit, K\spadesuit, Q\heartsuit, Q\diamondsuit, Q\clubsuit, Q\spadesuit,$ $J\heartsuit, J\diamondsuit, J\clubsuit, J\spadesuit\}$.

    (c) $B \cap C' = \{A\heartsuit, 2\heartsuit, \ldots, 10\heartsuit, A\diamondsuit, 2\diamondsuit, \ldots, 10\diamondsuit\}$.

    (d) $B' \cap C = \{K\clubsuit, K\spadesuit, Q\clubsuit, Q\spadesuit, J\clubsuit, J\spadesuit\}$
        $A \cup (B' \cap C) = \{A\heartsuit, A\diamondsuit, A\clubsuit, A\spadesuit, K\clubsuit, K\spadesuit, Q\clubsuit, Q\spadesuit, J\clubsuit, J\spadesuit\}$.

1.3.11   Let the event O be an on time repair and let the event S be a satisfactory repair. It is known that $P(O \cap S) = 0.26$, $P(O) = 0.74$ and $P(S) = 0.41$.

We want to find $P(O' \cap S')$.

Since the event $O' \cap S'$ can be written $(O \cup S)'$ it follows that

$P(O' \cap S') = 1 - P(O \cup S) = 1 - (P(O) + P(S) - P(O \cap S))$

$= 1 - (0.74 + 0.41 - 0.26) = 0.11$.

1.3.12   Let the event R be that a red ball is chosen and let the event S be that a shiny ball is chosen.

It is known that $P(R \cap S) = \frac{55}{200}$, $P(S) = \frac{91}{200}$ and $P(R) = \frac{79}{200}$.

Hence the probability that the chosen ball is either shiny or red is

$$P(R \cup S) = P(R) + P(S) - P(R \cap S) = \frac{79}{200} + \frac{91}{200} - \frac{55}{200} = \frac{115}{200} = 0.575.$$

The probability of a dull blue ball is $P(R' \cap S') = 1 - P(R \cup S)$

$= 1 - 0.575 = 0.425$.

1.4 Conditional Probability

1.4.1   See Figure 1.55.

(a) $P(A \mid B) = \frac{P(A \cap B)}{P(B)} = \frac{0.02 + 0.05 + 0.01}{0.02 + 0.05 + 0.01 + 0.11 + 0.08 + 0.06 + 0.13} = 0.1739.$

(b) $P(C \mid A) = \frac{P(A \cap C)}{P(A)} = \frac{0.02 + 0.05 + 0.08 + 0.04}{0.02 + 0.05 + 0.08 + 0.04 + 0.018 + 0.07 + 0.05} = 0.59375.$

(c) $P(B \mid A \cap B) = \frac{P(B \cap (A \cap B))}{P(A \cap B)} = \frac{P(A \cap B)}{P(A \cap B)} = 1.$

(d) $P(B \mid A \cup B) = \frac{P(B \cap (A \cup B))}{P(A \cup B)} = \frac{P(B)}{P(A \cup B)} = \frac{0.46}{0.46 + 0.32 - 0.08} = 0.657.$

(e) $P(A \mid A \cup B \cup C) = \frac{P(A \cap (A \cup B \cup C))}{P(A \cup B \cup C)} = \frac{P(A)}{P(A \cup B \cup C)} = \frac{0.32}{1 - 0.04 - 0.05 - 0.03}$
$= 0.3636.$

(f) $P(A \cap B \mid A \cup B) = \frac{P((A \cap B) \cap (A \cup B))}{P(A \cup B)} = \frac{P(A \cap B)}{P(A \cup B)} = \frac{0.08}{0.7} = 0.1143.$

1.4.2 $A = \{1, 2, 3, 5\}$ and $P(A) = \frac{4}{6} = \frac{2}{3}.$

$P(5 \mid A) = \frac{P(5 \cap A)}{P(A)} = \frac{P(5)}{P(A)} = \frac{\left(\frac{1}{6}\right)}{\left(\frac{2}{3}\right)} = \frac{1}{4}.$

$P(6 \mid A) = \frac{P(6 \cap A)}{P(A)} = \frac{P(\emptyset)}{P(A)} = 0.$

$P(A \mid 5) = \frac{P(A \cap 5)}{P(5)} = \frac{P(5)}{P(5)} = 1.$

1.4.3   (a) $P(A\heartsuit \mid \text{red suit}) = \frac{P(A\heartsuit \cap \text{red suit})}{P(\text{red suit})} = \frac{P(A\heartsuit)}{P(\text{red suit})} = \frac{\left(\frac{1}{52}\right)}{\left(\frac{26}{52}\right)} = \frac{1}{26}.$

(b) $P(\text{heart} \mid \text{red suit}) = \frac{P(\text{heart} \cap \text{red suit})}{P(\text{red suit})} = \frac{P(\text{heart})}{P(\text{red suit})} = \frac{\left(\frac{13}{52}\right)}{\left(\frac{26}{52}\right)} = \frac{13}{26} = \frac{1}{2}.$

(c) $P(\text{red suit} \mid \text{heart}) = \frac{P(\text{red suit} \cap \text{heart})}{P(\text{heart})} = \frac{P(\text{heart})}{P(\text{heart})} = 1.$

(d) $P(\text{heart} \mid \text{black suit}) = \frac{P(\text{heart} \cap \text{black suit})}{P(\text{black suit})} = \frac{P(\emptyset)}{P(\text{black suit})} = 0.$

(e) $P(\text{King} \mid \text{red suit}) = \frac{P(\text{King} \cap \text{red suit})}{P(\text{red suit})} = \frac{P(K\heartsuit, K\diamondsuit)}{P(\text{red suit})} = \frac{\left(\frac{2}{52}\right)}{\left(\frac{26}{52}\right)} = \frac{2}{26} = \frac{1}{13}.$

(f) $P(\text{King} \mid \text{red picture card}) = \frac{P(\text{King} \cap \text{red picture card})}{P(\text{red picture card})} = \frac{P(K\heartsuit, K\diamondsuit)}{P(\text{red picture card})}$
$= \frac{\left(\frac{2}{52}\right)}{\left(\frac{6}{52}\right)} = \frac{2}{6} = \frac{1}{3}.$

1.4.4 $P(A)$ is smaller than $P(A \mid B)$. Event B is a necessary condition for event A and so conditioning on event B increases the probability of event A.

1.4.5   There are 54 blue balls and so there are $150 - 54 = 96$ red balls. Also, there are 36 shiny, red balls and so there are $96 - 36 = 60$ dull, red balls.

$$P(\text{shiny} \mid \text{red}) \;=\; \frac{P(\text{shiny} \cap \text{red})}{P(\text{red})} \;=\; \frac{\left(\frac{36}{150}\right)}{\left(\frac{96}{150}\right)} \;=\; \frac{36}{96} \;=\; \frac{3}{8}.$$

$$P(\text{dull} \mid \text{red}) \;=\; \frac{P(\text{dull} \cap \text{red})}{P(\text{red})} \;=\; \frac{\left(\frac{60}{150}\right)}{\left(\frac{96}{150}\right)} \;=\; \frac{60}{96} \;=\; \frac{5}{8}.$$

1.4.6   Let the event O be an on time repair and let the event S be a satisfactory repair. It is known that $P(S \mid O) = 0.85$ and $P(O) = 0.77$.
The question asks for $P(O \cap S)$ which is
$$P(O \cap S) = P(S \mid O) \times P(O) = 0.85 \times 0.77 = 0.6545.$$

1.4.7   (a)   Probably increases although in some parts of the world it may decrease.

        (b)   Increases since there are proportionally more black haired people among brown eyed people than there are in the general population.

        (c)   Remains unchanged.

        (d)   Increases.

1.4.8   Over a four year period including one leap year, the number of days is $(3 \times 365) + 366 = 1{,}461$.

Of these $4 \times 12 = 48$ occur on the first day of a month and so the probability that a birthday falls on the first day of a month is $\frac{48}{1{,}461} = 0.0329$.

Of these 1,461 days $4 \times 31 = 124$ occur in March of which 4 days are March 1st. Consequently, the probability that a birthday falls on March 1st. conditional that it is in March is $\frac{4}{124} = \frac{1}{31} = 0.0323$.

Of these 1,461 days $(3 \times 28) + 29 = 113$ occur in February of which 4 days are February 1st. Consequently, the probability that a birthday falls on February 1st. conditional that it is in February is $\frac{4}{113} = 0.0354$.

1.4.9   (a)   Let A be the event that *Type I battery lasts longest* consisting of the outcomes $\{(\text{III, II, I}), (\text{II, III, I})\}$.
Let B be the event that *Type I battery does not fail first* consisting of the outcomes $\{(\text{III,II,I}), (\text{II,III,I}), (\text{II,I,III}), (\text{III,I,II})\}$.
The event $A \cap B = \{(\text{III,II,I}), (\text{II,III,I})\}$ is the same as event A.

        Hence $P(A \mid B) \;=\; \frac{P(A \cap B)}{P(B)} \;=\; \frac{0.39 + 0.03}{0.39 + 0.03 + 0.24 + 0.16} \;=\; 0.512.$

        (b)   Let C be the event that *Type II battery fails first* consisting of the outcomes $\{(\text{II,I,III}), (\text{II,III,I})\}$.
Thus, $A \cap C = \{(\text{II, III, I})\}$ and hence

$$P(A \mid C) \; = \; \tfrac{P(A \cap C)}{P(C)} \; = \; \tfrac{0.39}{0.39 \, + \, 0.24} \; = \; 0.619.$$

(c) Let D be the event that *Type II battery lasts longest* consisting of the outcomes
{(I,III,II), (III,I,II)}.
Thus, $A \cap D = \emptyset$ and hence

$$P(A \mid D) \; = \; \tfrac{P(A \cap D)}{P(D)} \; = \; 0.$$

(d) Let E be the event that *Type II battery does not fail first* consisting of the outcomes {(I,III,II), (I,II,III), (III,II,I), (III,I,II)}.
Therefore $A \cap E = \{(\text{III,II,I})\}$ and hence

$$P(A \mid E) \; = \; \tfrac{P(A \cap E)}{P(E)} \; = \; \tfrac{0.03}{0.07 + 0.11 + 0.03 + 0.16} \; = \; 0.081.$$

1.4.10   See Figure 1.25 in text.

(a) Let A be the event *both lines at full capacity* consisting of the outcome {(F,F)}.
Let B be the event *neither line is shut down* consisting of the outcomes
{(P,P), (P,F), (F,P), (F,F)}.
Therefore $A \cap B = \{(\text{F,F})\}$ and hence

$$P(A \mid B) \; = \; \tfrac{P(A \cap B)}{P(B)} \; = \; \tfrac{0.19}{(0.14 + 0.2 + 0.21 + 0.19)} \; = \; 0.257.$$

(b) Let C be the event *at least one line at full capacity* consisting of the outcomes
{(F,P), (F,S), (F,F), (S,F), (P,F)}.
Then $C \cap B = \{(\text{F,P}), (\text{F, F}), (\text{P,F})\}$ and hence

$$P(C \mid B) \; = \; \tfrac{P(C \cap B)}{P(B)} \; = \; \tfrac{0.21 \, + \, 0.19 \, + \, 0.2}{0.74} \; = \; 0.811.$$

(c) Let D be the event that *one line is at full capacity* consisting of the outcomes
{(F,P), (F,S), (P,F), (S,F)}.
Let E be the event *one line is shut down* consisting of the outcomes
{(S,P), (S,F), (P,S), (F,S)}.
Then $D \cap E = \{(\text{F,S}), (\text{S,F})\}$ and hence

$$P(D \mid E) \; = \; \tfrac{P(D \cap E)}{P(E)} \; = \; \tfrac{0.06 + 0.05}{0.06 + 0.05 + 0.07 + 0.06} \; = \; 0.458.$$

(d) Let G be the event that *neither line is at full capacity* consisting of the outcomes
{(S,S), (S,P), (P,S), (P,P)}.
Let H be the event that *at least one line is at partial capacity* consisting of the outcomes {(S,P), (P,S), (P,P), (P,F), (F,P)}.
Then $F \cap G = \{(\text{S,P}), (\text{P,S}), (\text{P,P})\}$ and hence

$$P(F \mid G) \; = \; \tfrac{P(F \cap G)}{P(G)} \; = \; \tfrac{0.06 \, + \, 0.07 \, + \, 0.14}{0.06 \, + \, 0.07 \, + \, 0.14 \, + \, 0.2 \, + \, 0.21} = 0.397.$$

1.5 Probabilities of Event Intersections

1.5.1 (a) P(both cards are picture cards) $= \frac{12}{52} \times \frac{11}{51} = \frac{132}{2652}$.

(b) P(both cards are from red suits) $= \frac{26}{52} \times \frac{25}{51} = \frac{650}{2652}$.

(c) P(one card is from a red suit and one is from black suit) =
(P(first card is red) $\times$ P(2nd card is black | 1st card is red))
+ (P(first card is black) $\times$ P(2nd card is red | 1st card is black))
$= \left(\frac{26}{52} \times \frac{26}{51} \right) + \left(\frac{26}{52} \times \frac{26}{51} \right) = \frac{676}{2652} \times 2 = \frac{26}{51}$.

1.5.2 (a) P(both cards are picture cards) $= \frac{12}{52} \times \frac{12}{52} = \frac{9}{169}$.
Probability increases with replacement.

(b) P(both cards are from red suits) $= \frac{26}{52} \times \frac{26}{52} = \frac{1}{4}$.
Probability increases with replacement.

(c) P(one card is from a red suit and one is from black suit) =
(P(first card is red) $\times$ P(2nd card is black | 1st card is red))
+ (P(first card is black) $\times$ P(2nd card is red | 1st card is black))
$= \left(\frac{26}{52} \times \frac{26}{52} \right) + \left(\frac{26}{52} \times \frac{26}{52} \right) = \frac{1}{2}$.
Probability decreases with replacement.

1.5.3 (a) No, they are not independent.
Notice that $P((ii)) = \frac{3}{13} \neq P((ii) \mid (i)) = \frac{11}{51}$.

(b) Yes, they are independent.
Notice that $P((i) \cup (ii)) = P((i)) \times P((ii))$ since $P((i)) = \frac{1}{4}$, $P((ii)) = \frac{3}{13}$ and
$P((i) \cup (ii)) = P(first\ card\ a\ heart\ picture \cup (ii))$
$+ P(first\ card\ a\ heart\ but\ not\ a\ picture \cup (ii))$
$= \left(\frac{3}{52} \times \frac{11}{51} \right) + \left(\frac{10}{52} \times \frac{12}{51} \right) = \frac{153}{2652} = \frac{3}{52}$.

(c) No, they are not independent.
Notice that $P((ii)) = \frac{1}{2} \neq P((ii) \mid (i)) = \frac{25}{51}$.

(d) Yes, they are independent. Similar to part (b).

(e) No, they are not independent.

1.5.4 P(all four cards are hearts) = P(1st card is a heart) $\times$ P(2nd card is a heart | 1st card is a heart) $\times$ P(3rd card is a heart | 1st and 2nd cards are hearts) $\times$ P(4th card is a heart | 1st, 2nd and 3rd cards are hearts) $= \frac{13}{52} \times \frac{12}{51} \times \frac{11}{50} \times \frac{10}{49} = 0.00264$.

P(all 4 cards from red suits) = P(1st card from red suit) $\times$ P(2nd card is from red suit | 1st card is from red suit) $\times$ P(3rd card is from red suit | 1st and 2nd cards are from red suits) $\times$ P(4th card is from red suit | 1st, 2nd and 3rd cards are from red suits) $= \frac{26}{52} \times \frac{25}{51} \times \frac{24}{50} \times \frac{23}{49} = 0.055$.

P(all 4 cards from different suits) = P(1st card from any suit) $\times$ P(2nd card not from suit of 1st card) $\times$ P(3rd card not from suit of 1st or 2nd cards) $\times$ P(4th card not from suit of 1st, 2nd, or 3rd cards) $= 1 \times \frac{39}{51} \times \frac{26}{50} \times \frac{13}{49} = 0.105$.

1.5.5   P(all 4 cards are hearts) $= (\frac{13}{52})^4 = \frac{1}{256}$.
        Probability increases with replacement.

        P(all 4 cards are from red suits) $= (\frac{26}{52})^4 = \frac{1}{16}$.
        Probability increases with replacement.

        P(all 4 cards from different suits) $= 1 \times \frac{39}{52} \times \frac{26}{52} \times \frac{13}{52} = \frac{3}{32}$.
        Probability decreases with replacement.

1.5.6   The events A and B are independent so that $P(A \mid B) = P(A)$, $P(B \mid A) = P(B)$,
        and $P(A \cap B) = P(A)P(B)$. To show that two events are independent, it must be
        shown that one of the above condition holds.

        (a) Recall that $P(A \cap B) + P(A \cap B') = P(A)$ and $P(B) + P(B') = 1$.
             Then
             $P(A \mid B') = \frac{P(A \cap B')}{P(B')} = \frac{P(A) - P(A \cap B)}{1 - P(B)} = \frac{P(A) - P(A)P(B)}{1 - P(B)} = \frac{P(A)(1 - P(B))}{1 - P(B)}$
             $= P(A)$.

        (b) Similar to part (a).

        (c) Now $P(A' \cap B') + P(A' \cap B) = P(A')$ so that
             $P(A' \cap B') = P(A) - P(A' \cap B) = P(A) - P(A')P(B)$
             since the events A' and B are independent.
             Thus $P(A' \cap B') = P(A)(1 - P(B)) = P(A')P(B')$.

1.5.7   The only way a message will not get through the network is if both branches are
        closed at the same time. The branches are independent since the switches operate
        independently of each other. Therefore,
        P(message gets through network)
        $= 1 - $ P(message cannot get through top branch or bottom branch)
        $= 1 - ($ P(message cannot get through top branch)
        $\times$ P(message cannot get through bottom branch)).

        P(message gets through top branch) = P(switch 1 open $\cap$ switch 2 open)
        = P(switch 1 open) $\times$ P(switch 2 open)
        since the switches are operate independently of each other.
        This is $0.88 \times 0.92 = 0.8096$.
        Therefore, P(message cannot get through top branch) =
        $1 - $ P(message gets through top branch) $= 1 - 0.8096 = 0.1904$.

        P(message cannot get through bottom branch) = P(switch 3 closed) $= 1 - 0.9 = 0.1$.

        Therefore, P(message gets through network) $= 1 - (0.1 \times 0.1904) = 0.98096$.

1.5.8   Given the birthday of the first person, the second person has a different birthday
        with a probability $\frac{364}{365}$. The third person has a different birthday from the first two

people with a probability $\frac{363}{365}$, and so the probability that all three people have different birthdays is $1 \times \frac{364}{365} \times \frac{363}{365}$.

Continuing in this manner the probability that n people all have different birthdays is therefore $\frac{364}{365} \times \frac{363}{365} \times \frac{362}{365} \times \ldots \times \frac{366-n}{365}$.

P(at least 2 people out of n share the same birthday)
= $1 - \text{P}(n$ people all have different birthdays)
= $1 - \left(\frac{364}{365} \times \frac{363}{365} \times \ldots \frac{366-n}{365} \right)$.

This probability is equal to 0.117 for $n = 10$, is equal to 0.253 for $n = 15$, is equal to 0.411 for $n = 20$, is equal to 0.569 for $n = 25$, is equal to 0.706 for $n = 30$, and is equal to 0.814 for $n = 35$. The smallest values of n for which the probability is greater than 0.5 is $n = 23$. In these calculations it has been assumed that birthdays are equally likely to occur on any day of the year, although in practice seasonal variations may be observed in the number of births.

1.5.9   P(no broken bulbs) $= \frac{83}{100} \times \frac{82}{99} \times \frac{81}{98} = 0.5682$.

P(one broken bulb) = P(broken, not broken, not broken)
+ P(not broken, broken, not broken) +P(not broken, not broken, broken)
= $\left(\frac{17}{100} \times \frac{83}{99} \times \frac{82}{98} \right) + \left(\frac{83}{100} \times \frac{17}{99} \times \frac{82}{98} \right) + \left(\frac{83}{100} \times \frac{82}{99} \times \frac{17}{98} \right) = 0.3578$.

P(no more than one broken bulb in the sample)
= P(no broken bulbs) + P(one broken bulb) = $0.5682 + 0.3578 = 0.9260$.

1.5.10   P(no broken bulbs) $= \frac{83}{100} \times \frac{83}{100} \times \frac{83}{100} = 0.5718$

P(one broken bulb) = P(broken, not broken, not broken)
+ P(not broken, broken, not broken) +P(not broken, not broken, broken)
= $\left(\frac{17}{100} \times \frac{83}{100} \times \frac{83}{100} \right) + \left(\frac{83}{100} \times \frac{17}{100} \times \frac{83}{100} \right) + \left(\frac{83}{100} \times \frac{83}{100} \times \frac{17}{100} \right) = 0.3513$.

P(no more than one broken bulb in the sample)
= P(no broken bulbs) + P(one broken bulb) = $0.5718 + 0.3513 = 0.9231$.

The probability of finding no broken bulbs increases with replacement, but the probability of finding no more than one broken bulb decreases with replacement.

1.5.11   P(drawing 2 green balls)
= P(1st ball is green) $\times$ P(2nd ball is green | 1st ball is green)
= $\frac{72}{169} \times \frac{71}{168} = 0.180$.

P(two balls same color)
= P(two red balls) + P(two blue balls) + P(two green balls)
= $\left(\frac{43}{169} \times \frac{42}{168} \right) + \left(\frac{54}{169} \times \frac{53}{168} \right) + \left(\frac{72}{169} \times \frac{71}{168} \right) = 0.344$.

P(two balls different colors) = 1 − P(two balls same color) = 1 − 0.344 = 0.656.

1.5.12   P(drawing 2 green balls) = $\frac{72}{169} \times \frac{72}{169}$ = 0.182.

P(two balls same color)
= P(two red balls) + P(two blue balls) + P(two green balls)
= $\left(\frac{43}{169} \times \frac{43}{169} \right) + \left(\frac{54}{169} \times \frac{54}{169} \right) + \left(\frac{72}{169} \times \frac{72}{169} \right)$ = 0.348.

P(two balls different colors) = 1 − P(two balls same color) = 1 − 0.348 = 0.652.

The probability that the two balls are green increases with replacement while the probability of drawing two balls of different colors decreases with replacement.

1.5.13   P(same result on both throws) = P(both heads) + P(both tails)
= $p^2 + (1 - p)^2 = 2p^2 - 2p + 1 = 2(p - 0.5)^2 + 0.5$ which is minimized when $p = 0.5$ (a fair coin).

1.6 Posterior Probabilities

1.6.1   (a) The following information is given: $P(\text{disease}) = 0.01$, $P(\text{no disease}) = 0.99$, $P(\text{positive blood test} \mid \text{disease}) = 0.97$, and $P(\text{positive blood test} \mid \text{no disease}) = 0.06$.

Hence, $P(\text{positive blood test}) = (P(\text{positive blood test} \mid \text{disease}) \times P(\text{disease}))$
$+ (P(\text{positive blood test} \mid \text{no disease}) \times P(\text{no disease}))$
$= (0.97 \times 0.01) + (0.06 \times 0.99) = 0.0691$.

(b) $P(\text{disease} \mid \text{positive blood test}) = \dfrac{P(\text{positive blood test} \cap \text{disease})}{P(\text{positive blood test})}$

$= \dfrac{P(\text{positive blood test} \mid \text{disease}) \times P(\text{disease})}{P(\text{positive blood test})} = \dfrac{0.97 \times 0.01}{0.0691} = 0.1404$.

(c) $P(\text{no disease} \mid \text{negative blood test}) = \dfrac{P(\text{no disease} \cap \text{negative blood test})}{P(\text{negative blood test})}$

$= \dfrac{P(\text{negative blood test} \mid \text{no disease}) \times P(\text{no disease})}{1 - P(\text{positive blood test})} = \dfrac{(1 - 0.06) \times 0.99}{(1 - 0.0691)} = 0.9997$.

1.6.2   (a) $P(\text{red}) = (P(\text{red} \mid \text{bag 1}) \times P(\text{bag 1})) + (P(\text{red} \mid \text{bag 2}) \times P(\text{bag 2}))$
$+ (P(\text{red} \mid \text{bag 3}) \times P(\text{bag 3}))$
$= \left(\frac{1}{3} \times \frac{3}{10}\right) + \left(\frac{1}{3} \times \frac{8}{12}\right) + \left(\frac{1}{3} \times \frac{5}{16}\right) = 0.426$.

(b) $P(\text{blue}) = 1 - P(\text{red}) = 1 - 0.426 = 0.574$.

(c) $P(\text{red ball from bag 2}) = P(\text{bag 2}) \times P(\text{red ball} \mid \text{bag 2}) = \frac{1}{3} \times \frac{8}{12} = \frac{2}{9}$.

$P(\text{bag 1} \mid \text{red ball}) = \dfrac{P(\text{bag 1} \cap \text{red ball})}{P(\text{red ball})} = \dfrac{P(\text{bag 1}) \times P(\text{red ball} \mid \text{bag 1})}{P(\text{red ball})}$
$= \dfrac{\frac{1}{3} \times \frac{3}{10}}{0.426} = 0.235$.

$P(\text{bag 2} \mid \text{blue ball}) = \dfrac{P(\text{bag 2} \cap \text{blue ball})}{P(\text{blue ball})} = \dfrac{P(\text{bag 2}) \times P(\text{blue ball} \mid \text{bag 1})}{P(\text{blue ball})}$
$= \dfrac{\frac{1}{3} \times \frac{4}{12}}{0.574} = 0.194$.

1.6.3   (a) $P(\text{Section I}) = \frac{55}{100}$.

(b) $P(\text{grade is A})$
$= (P(A \mid \text{Section I}) \times P(\text{Section I})) + (P(A \mid \text{Section II}) \times P(\text{Section II}))$
$= \left(\frac{10}{55} \times \frac{55}{100}\right) + \left(\frac{11}{45} \times \frac{45}{100}\right) = \frac{21}{100}$.

(c) $P(A \mid \text{Section I}) = \frac{10}{55}$.

(d) $P(\text{Section I} \mid A) = \dfrac{P(A \cap \text{Section I})}{P(A)} = \dfrac{P(\text{Section I}) \times P(A \mid \text{Section I})}{P(A)}$
$= \dfrac{\frac{55}{100} \times \frac{10}{55}}{\frac{21}{100}} = \frac{10}{21}$.

1.6.4   The following information is given: $P(\text{Species 1}) = 0.45$, $P(\text{Species 2}) = 0.38$, $P(\text{Species 3}) = 0.17$, $P(\text{Tagged} \mid \text{Species 1}) = 0.10$, $P(\text{Tagged} \mid \text{Species 2}) = 0.15$, $P(\text{Tagged} \mid \text{Species 3}) = 0.50$.

Therefore, P(Tagged) = (P(Tagged | Species 1) × P(Species 1))
+ (P(Tagged | Species 2) × P(Species 2)) + (P(Tagged | Species 3) × P(Species 3))
= (0.10 × 0.45) + (0.15 × 0.38) + (0.50 × 0.17) = 0.187.

$$P(\text{Species 1} \mid \text{Tagged}) = \frac{P(\text{Tagged} \cap \text{Species 1})}{P(\text{Tagged})}$$

$$= \frac{P(\text{Species 1}) \times P(\text{Tagged} \mid \text{Species 1})}{P(\text{Tagged})} = \frac{0.45 \times 0.10}{0.187} = 0.2406.$$

$$P(\text{Species 2} \mid \text{Tagged}) = \frac{P(\text{Tagged} \cap \text{Species 2})}{P(\text{Tagged})}$$

$$= \frac{P(\text{Species 2}) \times P(\text{Tagged} \mid \text{Species 2})}{P(\text{Tagged})} = \frac{0.38 \times 0.15}{0.187} = 0.3048.$$

$$P(\text{Species 3} \mid \text{Tagged}) = \frac{P(\text{Tagged} \cap \text{Species 3})}{P(\text{Tagged})}$$

$$= \frac{P(\text{Species 3}) \times P(\text{Tagged} \mid \text{Species 3})}{P(\text{Tagged})} = \frac{0.17 \times 0.50}{0.187} = 0.4545.$$

1.7 Counting Techniques

1.7.1   (a) $7! = 7 \times 6 \times 5 \times 4 \times 3 \times 2 \times 1 = 5{,}040$.

          (b) $8! = 8 \times 7! = 40{,}320$.

          (c) $4! = 4 \times 3 \times 2 \times 1 = 24$.

          (d) $13! = 13 \times 12 \times 11 \times \ldots \times 1 = 6{,}227{,}020{,}800$.

1.7.2   (a) $P_2^7 = \frac{7!}{(7-2)!} = 7 \times 6 = 42$.

          (b) $P_5^9 = \frac{9!}{(9-5)!} = 9 \times 8 \times 7 \times 6 \times 5 = 15{,}120$.

          (c) $P_2^5 = \frac{5!}{(5-2)!} = 5 \times 4 = 20$.

          (d) $P_4^{17} = \frac{17!}{(17-4)!} = 17 \times 16 \times 15 \times 14 = 57{,}120$.

1.7.3   (a) $C_2^6 = \frac{6!}{(6-2)! \times 2!} = \frac{6 \times 5}{2} = 15$.

          (b) $C_4^8 = \frac{8!}{(8-4)! \times 4!} = \frac{8 \times 7 \times 6 \times 5}{24} = 70$.

          (c) $C_2^5 = \frac{5!}{(5-2)! \times 2!} = \frac{5 \times 4}{2} = 10$.

          (d) $C_6^{14} = \frac{14!}{(14-6)! \times 6!} = 3{,}003$.

1.7.4   Number of full meals $= 5 \times 3 \times 7 \times 6 \times 8 = 5{,}040$.

       Number of meals with just soup or appetizer $= (5+3) \times 7 \times 6 \times 8 = 2{,}688$.

1.7.5   Number of experimental configurations $= 3 \times 4 \times 2 = 24$.

1.7.6   (a) Define the notation (2,3,1,4) to represent the result that the player who finished 1st in tournament 1 finished 2nd in tournament 2, the player who finished 2nd in tournament 1 finished 3rd in tournament 2, etc. The result (1,2,3,4) then indicates that each competitor received the same ranking in both tournaments. Altogether there are $4! = 24$ different results, each equally likely, and so this single result has a probability of $\frac{1}{24}$.

          (b) The results where no player receives the same ranking in the two tournaments are (2,1,4,3), (2,3,4,1), (2,4,1,3), (3,1,4,2), (3,4,1,2) (3,4,2,1), (4,1,2,3), (4,3,1,2) and (4,3,2,1). There are nine of these and so the required probability is $\frac{9}{24} = \frac{3}{8}$.

1.7.7   Number of rankings of top 5 $= P_5^{20} = \frac{20!}{15!} = 20 \times 19 \times 18 \times 17 \times 16 = 1{,}860{,}480$.

       Number of combinations of top 5 $= C_5^{20} = \frac{20!}{15! \times 5!} = 15{,}504$.

1.7.8   (a) $C_3^{100} = \frac{100!}{97! \times 3!} = \frac{100 \times 99 \times 98}{6} = 161{,}700$.

(b) $C_3^{83} = \frac{83!}{80! \times 3!} = \frac{83 \times 82 \times 81}{6} = 91,881.$

(c) P(no broken lightbulbs) $= \frac{91,881}{161,700} = 0.568.$

(d) $17 \times C_2^{83} = 17 \times \frac{83 \times 82}{2} = 57,851.$

(e) Number of samples with 0 or 1 broken bulbs $= 91,881 + 57,851 = 149,732.$
    P(sample contains no more than 1 broken bulb) $= \frac{149,732}{161,700} = 0.926.$

1.7.9 $C_k^{n-1} + C_{k-1}^{n-1} = \frac{(n-1)!}{k!(n-1-k)!} + \frac{(n-1)!}{(k-1)!(n-k)!} = \frac{n!}{k!(n-k)!} \left(\frac{n-k}{n} + \frac{k}{n} \right) = \frac{n!}{k!(n-k)!} = C_k^n.$

This relationship can be interpreted in the following manner. C_k^n is the number of ways k balls can be selected from n balls. Suppose that one ball is red while the remaining $n - 1$ balls are blue. Either all k balls selected are blue or one of the selected balls is red, and C_k^{n-1} is the number of ways k blue balls can be selected while C_{k-1}^{n-1} is the number of ways of selecting the one red ball and $k - 1$ blue balls.

1.7.10   (a) Number of possible 5 card hands $= C_5^{52} = \frac{52!}{47! \times 5!} = 2,598,960.$

(b) Number of ways to get a hand of 5 hearts $= C_5^{13} = \frac{13!}{8! \times 5!} = 1,287.$

(c) Number of ways to get a flush $= 4 \times C_5^{13} = 4 \times 1,287 = 5,148.$

(d) P(flush) $= \frac{5,148}{2,598,960} = 0.00198.$

(e) There are 48 choices for the fifth card in the hand and so the number of hands containing all four aces is 48.

(f) $13 \times 48 = 624.$

(g) P(hand has four cards of the same number or picture) $= \frac{624}{2,598,960} = 0.00024.$

1.7.11   The number of ways n objects can be arranged in a line is $n!$. If the line is made into a circle and rotations of the circle are considered to be indistinguishable, then there are n arrangements of the line corresponding to each arrangement of the circle. Consequently, there are $\frac{n!}{n} = (n - 1)!$ ways to order the objects in a circle.

1.7.12   Number of ways six people can sit in a line at cinema $= 6! = 720.$

Number of ways six people can sit at dinner table $= 5! = 120$ (see previous problem).

1.7.13   Consider 5 blocks, one block being Andrea and Scott and the other four blocks being the other four people. At the cinema these 5 blocks can be arranged in 5! ways, and then Andrea and Scott can be arranged in two different ways within their block so that the total number of seating arrangements is $2 \times 5! = 240$. The total number of seating arrangements at the dinner table is $2 \times 4! = 48$.

If Andrea refuses to sit next to Scott the number of seating arrangements can be obtained by subtraction. The total number of seating arrangements at the cinema is

720 − 240 = 480 and the total number of seating arrangements at the dinner table is 120 − 48 = 72.

1.7.15   (a) From the previous problem the answer is $\frac{12!}{3! \times 4! \times 5!} = 27{,}720$.

      (b) Suppose that the balls in (a) are labeled 1 to 12. Then the positions of the three red balls in the line (where the places in the line are labeled 1 to 12) can denote which balls in (a) are placed in the first box, the positions of the four blue balls in the line can denote which balls in (a) are placed in the second box, and the positions of the five green balls in the line can denote which balls in (a) are placed in the third box. Thus there is a one-to-one correspondence between the positioning of the colored balls in (b) and the arrangements of the balls in (a) so that the problems are identical.

1.7.16 $\frac{14!}{3! \times 4! \times 7!} = 120{,}120$.

1.7.17 $\frac{15!}{3! \times 3! \times 3! \times 3! \times 3!} = 168{,}168{,}000$.

1.8 Supplementary Problems

1.8.1 $S = \{A, B, C, D, F\}$

1.8.2   If the four contestants are labeled A, B, C, D and the notation (X,Y) is used to indicate that contestant X is the winner and contestant Y is the runner up, then the sample space is $S = \{(A,B), (A,C), (A,D), (B,A), (B,C), (B,D), (C,A), (C,B), (C,D), (D,A), (D,B), (D,C)\}$.

1.8.3   One way is to have the two team captains each toss the coin once. If one obtains a head and the other a tail, then the one with the head wins (this could just as well be done the other way around so that the one with the tail wins, as long as it is decided beforehand). If both captains obtain the same result, that is if there are two heads or two tails, then the procedure is repeated.

1.8.4   See Figure 1.10.
There are 36 equally likely outcomes, 16 of which have scores differing by no more than one.
Thus, P(scores on two dice differ by no more than one) $= \frac{16}{36} = \frac{4}{9}$.

1.8.5   Number of ways to pick a card $= 52$.
Number of ways to pick a diamond picture card $= 3$.
Thus, P(picking a diamond picture card) $= \frac{3}{52}$.

1.8.6   With replacement, P(drawing two hearts) $= \frac{13}{52} \times \frac{13}{52} = \frac{1}{16} = 0.0625$.

Without replacement, P(drawing two hearts) $= \frac{13}{52} \times \frac{12}{51} = \frac{3}{51} = 0.0588$.

The probability decreases without replacement.

1.8.7   A $= \{(1,1), (1,2), (1,3), (2,1), (2,2), (3,1)\}$.
B $= \{(1,1), (2,2), (3,3), (4,4), (5,5), (6,6)\}$.

(a) A $\cap$ B $= \{(1,1), (2,2)\}$ and P(A $\cap$ B) $= \frac{2}{36} = \frac{1}{18}$.

(b) A $\cup$ B $= \{(1,1), (1,2), (1,3), (2,1), (2,2), (3,1), (3,3), (4,4), (5,5), (6,6)\}$
and P(A $\cup$ B) $= \frac{10}{36} = \frac{5}{18}$.

(c) $A' \cup$ B $= \{(1,1), (1,4), (1,5), (1,6), (2,2), (2,3), (2,4), (2,5), (2,6), (3,2), (3,3), (3,4), (3,5), (3,6), (4,1), (4,2), (4,3), (4,4), (4,5), (4,6), (5,1), (5,2), (5,3), (5,4), (5,5), (5,6), (6,1), (6,2), (6,3), (6,4), (6,5), (6,6)\}$
and P($A' \cup$ B) $= \frac{32}{36} = \frac{8}{9}$.

1.8.8   See Figure 1.10. Let the notation (x, y) indicate that the score on the red die is x and the score on the blue die is y.

  (a) The event *the sum of the scores on the two dice is eight* consists of the outcomes $\{(2,6), (3,5), (4,4), (5,3), (6,2)\}$.

  Thus P(red die is 5 | sum of scores is 8) $= \dfrac{\text{P(red die is 5} \cap \text{sum of scores is 8)}}{\text{P(sum of scores is 8)}}$

  $= \dfrac{\left(\frac{1}{36}\right)}{\left(\frac{5}{36}\right)} = \frac{1}{5}$.

  (b) P(either score is 5 | sum of scores is 8) $= 2 \times \frac{1}{5} = \frac{2}{5}$.

  (c) The event *the score on either die is 5* consists of the 11 outcomes $\{(1,5), (2,5), (3,5), (4,5), (5,5), (6,5), (5,6), (5,4), (5,3), (5,2), (5,1)\}$.

  Thus P(sum of scores is 8 | either score is 5) $= \dfrac{\text{P(sum of scores is 8} \cap \text{either score is 5)}}{\text{P(either score is 5)}}$

  $= \dfrac{\left(\frac{2}{36}\right)}{\left(\frac{11}{36}\right)} = \frac{2}{11}$.

1.8.9   P(A) = P(either switch 1 or 4 is open or both)
  $= 1 - $ P(both switches 1 and 4 are closed) $= 1 - 0.15^2 = 0.9775$.

  P(B) = P(either switch 2 or 5 is open or both)
  $= 1 - $ P(both switches 2 and 5 are closed) $= 1 - 0.15^2 = 0.9775$.

  P(C) = P(switches 1 and 2 are both open) $= 0.85^2 = 0.7225$

  P(D) = P(switches 4 and 5 are both open) $= 0.85^2 = 0.7225$

  If E = C $\cup$ D then P(E) $= 1 - (\text{P}(C') \times \text{P}(D')) = 1 - (1 - 0.85^2)^2 = 0.923$.

  P(message gets through the network)
  $= (\text{P(switch 3 is open)} \times \text{P(A)} \times \text{P(B)}) + (\text{P(switch 3 closed)} \times \text{P(E)})$
  $= (0.85 \times (1 - 0.15^2)^2) + (0.15 \times (1 - (1 - 0.85^2)^2)) = 0.9506$.

1.8.10 The sample space for the experiment of two coin tosses consists of the equally likely outcomes $\{(H,H), (H,T), (T,H), (T,T)\}$. Three of these four outcomes contain at least one head so that P(at least one head in two coin tosses) $= \frac{3}{4}$.

  The sample space for four tosses of a coin consists of $2^4 = 16$ equally likely outcomes of which the following 11 outcomes contain at least two heads $\{(HHTT), (HTHT), (HTTH), (THHT), (THTH), (TTHH), (HHHT), (HHTH), (HTHH), (THHH), (HHHH)\}$.
  Therefore P(at least two heads in four coin tosses) $= \frac{11}{16}$ which is smaller than the previous probability.

1.8.11   (a) P(blue ball) = (P(bag 1) × P(blue ball | bag 1))
                 + (P(bag 2) × P(blue ball | bag 2)) + (P(bag 3) × P(blue ball | bag 3))
                 + (P(bag 4) × P(blue ball | bag 4))
                 $= \left(0.15 \times \frac{7}{16}\right) + \left(0.2 \times \frac{8}{18}\right) + \left(0.35 \times \frac{9}{19}\right) + \left(0.3 \times \frac{7}{11}\right) = 0.5112.$

         (b) $\text{P(bag 4 | green ball)} = \dfrac{\text{P(green ball} \cap \text{bag 4)}}{\text{P(green ball)}} = \dfrac{\text{P(bag 4)} \times \text{P(green ball | bag 4)}}{\text{P(greenball)}}$
                 $= \dfrac{0.3 \times 0}{\text{P(green ball)}} = 0.$

         (c) $\text{P(bag 1 | blue ball)} = \dfrac{\text{P(bag 1)} \times \text{P(blue ball | bag 1)}}{\text{P(blue ball)}} = \dfrac{0.15 \times \frac{7}{16}}{0.5112}$
                 $= \dfrac{0.0656}{0.5112} = 0.128.$

1.8.12   (a) $\mathcal{S} = \{1,\ 2,\ 3,\ 4,\ 5,\ 6,\ 10\}$

         (b) $\text{P(10)} = \text{P(score on die is 5)} \times \text{P(tails)} = \frac{1}{6} \times \frac{1}{2} = \frac{1}{12}.$

         (c) $\text{P(3)} = \text{P(score on die is 3)} \times \text{P(heads)} = \frac{1}{6} \times \frac{1}{2} = \frac{1}{12}.$

         (d) $\text{P(6)} = \text{P(score on die is 6)} + (\text{P(score on die is 3)} \times \text{P(tails)})$
                 $= \frac{1}{6} + (\frac{1}{6} \times \frac{1}{2}) = \frac{1}{4}.$

         (e) 0.

         (f) $\text{P(score on die is odd | 6 is recorded)} = \dfrac{\text{P(score on die is odd} \cap \text{6 is recorded)}}{\text{P(6 is recorded)}}$
                 $= \dfrac{\text{P(score on die is 3)} \times \text{P(tails)}}{\text{P(6 is recorded)}} = \dfrac{\left(\frac{1}{12}\right)}{\left(\frac{1}{4}\right)} = \frac{1}{3}.$

1.8.13 $5^4 = 625$
         $4^5 = 1{,}024$
         In this case $5^4 < 4^5$ and in general, $n_2^{n_1} < n_1^{n_2}$ when $3 \leq n_1 < n_2$.

1.8.14 $\dfrac{20!}{5! \times 5! \times 5! \times 5!} = 1.17 \times 10^{10}.$
         $\dfrac{20!}{4! \times 4! \times 4! \times 4 \times 4!} = 3.06 \times 10^{11}.$

1.8.15 $P(X = 0) = \frac{1}{4},\ P(X = 1) = \frac{1}{2},\ P(X = 2) = \frac{1}{4}.$
         $P(X = 0|\text{white}) = \frac{1}{8},\ P(X = 1|\text{white}) = \frac{1}{2},\ P(X = 2|\text{white}) = \frac{3}{8}.$
         $P(X = 0|\text{black}) = \frac{1}{2},\ P(X = 1|\text{black}) = \frac{1}{2},\ P(X = 2|\text{black}) = 0.$

Chapter 2

Random Variables

2.1 Discrete Random Variables

2.1.1   (a) $0.08 + 0.11 + 0.27 + 0.33 + P(X = 4) = 1 \quad \Rightarrow \quad P(X = 4) = 0.21$.

        (c) $F(0) = 0.08$, $F(1) = 0.19$, $F(2) = 0.46$, $F(3) = 0.79$, $F(4) = 1.00$.

2.1.2

x_i	-4	-1	0	2	3	7
p_i	0.21	0.11	0.07	0.29	0.13	0.19

2.1.3

x_i	1	2	3	4	5	6	8	9	10
p_i	$\frac{1}{36}$	$\frac{2}{36}$	$\frac{2}{36}$	$\frac{3}{36}$	$\frac{2}{36}$	$\frac{4}{36}$	$\frac{2}{36}$	$\frac{1}{36}$	$\frac{2}{36}$
$F(x_i)$	$\frac{1}{36}$	$\frac{3}{36}$	$\frac{5}{36}$	$\frac{8}{36}$	$\frac{10}{36}$	$\frac{14}{36}$	$\frac{16}{36}$	$\frac{17}{36}$	$\frac{19}{36}$

x_i	12	15	16	18	20	24	25	30	36
p_i	$\frac{4}{36}$	$\frac{2}{36}$	$\frac{1}{36}$	$\frac{2}{36}$	$\frac{2}{36}$	$\frac{2}{36}$	$\frac{1}{36}$	$\frac{2}{36}$	$\frac{1}{36}$
$F(x_i)$	$\frac{23}{36}$	$\frac{25}{36}$	$\frac{26}{36}$	$\frac{28}{36}$	$\frac{30}{36}$	$\frac{32}{36}$	$\frac{33}{36}$	$\frac{35}{36}$	1

2.1.4   (a)

x_i	0	1	2
p_i	0.5625	0.3750	0.0625

        (b)

x_i	0	1	2
$F(x_i)$	0.5625	0.9375	1.000

        (c) $x = 0$ is the most likely value.

Without replacement:

x_i	0	1	2
p_i	0.5588	0.3824	0.0588
$F(x_i)$	0.5588	0.9412	1.000

$x = 0$ is the most likely value.

2.1.5

x_i	-5	-4	-3	-2	-1	0	1	2	3	4	6	8	10	12
p_i	$\frac{1}{36}$	$\frac{1}{36}$	$\frac{2}{36}$	$\frac{2}{36}$	$\frac{3}{36}$	$\frac{3}{36}$	$\frac{2}{36}$	$\frac{5}{36}$	$\frac{1}{36}$	$\frac{4}{36}$	$\frac{3}{36}$	$\frac{3}{36}$	$\frac{3}{36}$	$\frac{3}{36}$
$F(x_i)$	$\frac{1}{36}$	$\frac{2}{36}$	$\frac{4}{36}$	$\frac{6}{36}$	$\frac{9}{36}$	$\frac{12}{36}$	$\frac{14}{36}$	$\frac{19}{36}$	$\frac{20}{36}$	$\frac{24}{36}$	$\frac{27}{36}$	$\frac{30}{36}$	$\frac{33}{36}$	1

2.1.6 (a)

x_i	-6	-4	-2	0	2	4	6
p_i	$\frac{1}{8}$	$\frac{1}{8}$	$\frac{1}{8}$	$\frac{2}{8}$	$\frac{1}{8}$	$\frac{1}{8}$	$\frac{1}{8}$

(b)

x_i	-6	-4	-2	0	2	4	6
$F(x_i)$	$\frac{1}{8}$	$\frac{2}{8}$	$\frac{3}{8}$	$\frac{5}{8}$	$\frac{6}{8}$	$\frac{7}{8}$	1

(c) $x = 0$ is the most likely value.

2.1.7 (a)

x_i	0	1	2	3	4	6	8	12
p_i	0.061	0.013	0.195	0.067	0.298	0.124	0.102	0.140

(b)

x_i	0	1	2	3	4	6	8	12
$F(x_i)$	0.061	0.074	0.269	0.336	0.634	0.758	0.860	1.000

(c) The most likely value is 4.

P(not shipped) = P(X $\leq$ 1) = 0.074.

2.1.8

x_i	-1	0	1	3	4	5
p_i	$\frac{1}{6}$	$\frac{1}{6}$	$\frac{1}{6}$	$\frac{1}{6}$	$\frac{1}{6}$	$\frac{1}{6}$
$F(x_i)$	$\frac{1}{6}$	$\frac{2}{6}$	$\frac{3}{6}$	$\frac{4}{6}$	$\frac{5}{6}$	1

2.1.9

x_i	1	2	3	4
p_i	$\frac{2}{5}$	$\frac{3}{10}$	$\frac{1}{5}$	$\frac{1}{10}$
$F(x_i)$	$\frac{2}{5}$	$\frac{7}{10}$	$\frac{9}{10}$	1

2.1.10 Since $\sum_{i=1}^{\infty} \frac{1}{i^2} = \frac{\pi^2}{6}$, $P(X = i) = \frac{6}{\pi^2 i^2}$ is a possible set of probability values.

However, since $\sum_{i=1}^{\infty} \frac{1}{i}$ diverges $P(X = i) = \frac{c}{i}$ is not a possible set of probability values.

2.2   Continuous Random Variables

2.2.1   (a) Continuous.

(b) Discrete.

(c) Continuous.

(d) Continuous.

(e) Discrete.

(f) This depends on what level of accuracy it is measured to. It can be considered to be either discrete or continuous.

2.2.2   (b)
$$\int_4^6 \frac{1}{x\ln(1.5)}\,dx = \frac{1}{\ln(1.5)} \times [\ln(x)]_4^6 = \frac{1}{\ln(1.5)} \times (\ln(6) - \ln(4)) = 1.0.$$

(c)
$$P(4.5 \le X \le 5.5) = \int_{4.5}^{5.5} \frac{1}{x\ln(1.5)}\,dx = \frac{1}{\ln(1.5)} \times [\ln(x)]_{4.5}^{5.5}$$

$$= \frac{1}{\ln(1.5)} \times (\ln(5.5) - \ln(4.5)) = 0.495.$$

(d)
$$F(x) = \int_4^x \frac{1}{y\ln(1.5)}\,dy = \frac{1}{\ln(1.5)} \times [\ln(y)]_4^x = \frac{1}{\ln(1.5)} \times (\ln(x) - \ln(4))$$
for $4 \le x \le 6$.

2.2.3   (a)
$$\int_{-2}^0 \left(\frac{15}{64} + \frac{x}{64}\right) dx = \frac{7}{16} \quad \text{and} \quad \int_0^3 \left(\frac{3}{8} + cx\right) dx = \frac{9}{8} + \frac{9c}{2}.$$

Then $\frac{7}{16} + \frac{9}{8} + \frac{9c}{2} = 1$ gives $c = -\frac{1}{8}$.

(b)
$$P(-1 \le X \le 1) = \int_{-1}^0 \left(\frac{15}{64} + \frac{x}{64}\right) dx + \int_0^1 \left(\frac{3}{8} - \frac{x}{8}\right) dx = 0.618.$$

(c)
$$F(x) = \int_{-2}^x \left(\frac{15}{64} + \frac{y}{64}\right) dy = \frac{x^2}{128} + \frac{30x}{128} + \frac{7}{16} \quad \text{for } -2 \le x \le 0.$$

$$F(x) = \frac{7}{16} + \int_0^x \left(\frac{3}{8} - \frac{y}{8}\right) dy = -\frac{x^2}{16} + \frac{3x}{8} + \frac{7}{16} \quad \text{for } 0 \le x \le 3.$$

2.2.4   (b) $P(X \le 2) = F(2) = \frac{1}{4}$

(c) $P(1 \le X \le 3) = F(3) - F(1) = \frac{9}{16} - \frac{1}{16} = \frac{1}{2}.$

(d)
$$f(x) = \frac{dF(x)}{dx} = \frac{x}{8} \quad \text{for } 0 \le x \le 4.$$

2.2.5   (a) $F(\infty) = 1$ gives $A = 1$. Then $F(0) = 0$ gives $1 + B = 0$ so $B = -1$ and $F(x) = 1 - e^{-x}$.

   (b) $P(2 \leq X \leq 3) = F(3) - F(2) = e^{-2} - e^{-3} = 0.0855$.

   (c) $$f(x) = \frac{dF(x)}{dx} = e^{-x} \quad \text{for} \ x \geq 0.$$

2.2.6   (a) $$\int_{0.125}^{0.5} A \left(0.5 - (x - 0.25)^2\right) dx = 1 \quad \Rightarrow \quad A = 5.5054.$$

   (b) $$F(x) = \int_{0.125}^{x} f(y)dy = 5.5054 \left(\frac{x}{2} - \frac{(x - 0.25)^3}{3} - 0.06315\right)$$
   for $0.125 \leq x \leq 0.5$.

   (c) $F(0.2) = 0.203$.

2.2.7   (a) $F(0) = A + B\ln(2) = 0$ and $F(10) = A + B\ln(32) = 1$ give $A = -0.25$ and $B = \frac{1}{\ln(16)} = 0.361$.

   (b) $P(X > 2) = 1 - F(2) = 0.5$.

   (c) $$f(x) = \frac{dF(x)}{dx} = \frac{1.08}{3x + 2} \quad \text{for} \ 0 \leq x \leq 10.$$

2.2.8   (a) $$\int_{0}^{10} A \left(e^{10-\theta} - 1\right) d\theta = 1 \quad \Rightarrow \quad A = (e^{10} - 11)^{-1} = 4.54 \times 10^{-5}.$$

   (b) $$F(\theta) = \int_{0}^{\theta} f(y)dy = \frac{e^{10} - \theta - e^{10-\theta}}{e^{10} - 11} \quad \text{for} \ 0 \leq \theta \leq 10.$$

   (c) $1 - F(8) = 0.0002$.

2.2.9   (a) $F(0) = 0$ and $F(50) = 1$ give $A = 1.0007$ and $B = -125.09$.
   (b) $P(X \leq 10) = F(10) = 0.964$.
   (c) $P(X \geq 30) = 1 - F(30) = 1 - 0.998 = 0.002$.
   (d) $$f(r) = \frac{dF(r)}{dr} = \frac{375.3}{(r + 5)^4} \quad \text{for} \ 0 \leq r \leq 50.$$

2.2.10   (a) $F(200) = 0.1$.
   (b) $F(700) - F(400) = 0.65$.

2.3   The Expectation of a Random Variable

2.3.1 $E(X) = (0 \times 0.08) + (1 \times 0.11) + (2 \times 0.27) + (3 \times 0.33) + (4 \times 0.21) = 2.48.$

2.3.2 $E(X) = \left(1 \times \frac{1}{36}\right) + \left(2 \times \frac{2}{36}\right) + \left(3 \times \frac{2}{36}\right) + \left(4 \times \frac{3}{36}\right) + \left(5 \times \frac{2}{36}\right) + \left(6 \times \frac{4}{36}\right) +$
$\left(8 \times \frac{2}{36}\right) + \left(9 \times \frac{1}{36}\right) + \left(10 \times \frac{2}{36}\right) + \left(12 \times \frac{4}{36}\right) + \left(15 \times \frac{2}{36}\right) + \left(16 \times \frac{1}{36}\right) + \left(18 \times \frac{2}{36}\right)$
$+ \left(20 \times \frac{2}{36}\right) + \left(24 \times \frac{2}{36}\right) + \left(25 \times \frac{1}{36}\right) + \left(30 \times \frac{2}{36}\right) + \left(36 \times \frac{1}{36}\right) = 12.25.$

2.3.3   With replacement $E(X) = (0 \times 0.5625) + (1 \times 0.3750) + (2 \times 0.0625) = 0.5.$

Without replacement $E(X) = (0 \times 0.5588) + (1 \times 0.3824) + (2 \times 0.0588) = 0.5.$

2.3.4 $E(X) = \left(1 \times \frac{2}{5}\right) + \left(2 \times \frac{3}{10}\right) + \left(3 \times \frac{1}{5}\right) + \left(4 \times \frac{1}{10}\right) = 2.$

2.3.5

x_i	2	3	4	5	6	7	8	9	10	15
p_i	$\frac{1}{13}$	$\frac{1}{13}$	$\frac{1}{13}$	$\frac{1}{13}$	$\frac{1}{13}$	$\frac{1}{13}$	$\frac{1}{13}$	$\frac{1}{13}$	$\frac{1}{13}$	$\frac{4}{13}$

$$E(X) = \left(2 \times \frac{1}{13}\right) + \left(3 \times \frac{1}{13}\right) + \left(4 \times \frac{1}{13}\right) + \left(5 \times \frac{1}{13}\right) + \left(6 \times \frac{1}{13}\right)$$

$$+ \left(7 \times \frac{1}{13}\right) + \left(8 \times \frac{1}{13}\right) + \left(9 \times \frac{1}{13}\right) + \left(10 \times \frac{1}{13}\right) + \left(15 \times \frac{4}{13}\right) = \$8.77.$$

If I paid $9 to play, my expected loss would be 23 cents.

2.3.6

x_i	1	2	3	4	5	6	7	8	9	10	11	12
p_i	$\frac{6}{72}$	$\frac{7}{72}$	$\frac{8}{72}$	$\frac{9}{72}$	$\frac{10}{72}$	$\frac{11}{72}$	$\frac{6}{72}$	$\frac{5}{72}$	$\frac{4}{72}$	$\frac{3}{72}$	$\frac{2}{72}$	$\frac{1}{72}$

$$E(X) = \left(1 \times \frac{6}{72}\right) + \left(2 \times \frac{6}{72}\right) + \left(3 \times \frac{6}{72}\right) + \left(4 \times \frac{6}{72}\right) + \left(5 \times \frac{6}{72}\right)$$

$$+ \left(6 \times \frac{6}{72}\right) + \left(7 \times \frac{6}{72}\right) + \left(8 \times \frac{6}{72}\right) + \left(9 \times \frac{6}{72}\right) + \left(10 \times \frac{6}{72}\right) + \left(11 \times \frac{6}{72}\right)$$

$$+ \left(12 \times \frac{6}{72}\right) = 5.25.$$

2.3.7 P(3 sixes are rolled) $= \frac{1}{6} \times \frac{1}{6} \times \frac{1}{6} = \frac{1}{216}$.

Therefore E(net winnings) $= (-\$1 \times \frac{215}{216}) + (\$499 \times \frac{1}{216}) = \$1.31$.

If you can play the game a large number of times then you should play the game as often as you can.

2.3.8 The expected net winnings will be negative. However, people play because the cost of a ticket is small and they imagine great benefit if they happy to become very rich.

2.3.9

x_i	0	1	2	3	4	5
p_i	0.1680	0.2816	0.2304	0.1664	0.1024	0.0512

E(payment) $= (0 \times 0.1680) + (1 \times 0.2816) + (2 \times 0.2304) + (3 \times 0.1664)$
$+ (4 \times 0.1024) + (5 \times 0.0512) = 1.9072$

E(winnings) $= \$2 - \$1.91 = \$0.09$.

The expected winnings increase to 9 cents per game. Increasing the probability of scoring a three reduces the expected value of the difference in the scores of the two dice.

2.3.10 (a) $$E(X) = \int_4^6 x \, \frac{1}{x \ln(1.5)} \, dx = 4.94.$$

(b) Solving $F(x) = 0.5$ gives $x = 4.90$.

2.3.11 (a) $$E(X) = \int_0^4 x \, \frac{x}{8} \, dx = 2.67.$$

(b) Solving $F(x) = 0.5$ gives $x = \sqrt{8} = 2.83$.

2.3.12 $$E(X) = \int_{0.125}^{0.5} x \, 5.5054(0.5 - (x - 0.25)^2) \, dx = 0.3095.$$

Solving $F(x) = 0.5$ gives $x = 0.3081$.

2.3.13 $$E(X) = \int_0^{10} \frac{\theta}{e^{10} - 11} (e^{10-\theta} - 1) \, d\theta = 0.9977.$$

Solving $F(\theta) = 0.5$ gives $\theta = 0.6927$.

2.3.14 $\qquad E(X) \; = \; \int_0^{50} \dfrac{375.3\, r}{(r+5)^4} \; dr \; = \; 2.44.$

Solving $F(r) = 0.5$ gives $r = 1.30$.

2.3.15 Let $f(x)$ be a probability density function symmetric about the point μ so that $f(\mu + x) = f(\mu - x)$. Then

$$E(X) \; = \; \int_{-\infty}^{\infty} xf(x)\,dx$$

which under the transformation $x = \mu + y$ gives

$$E(X) \; = \; \int_{-\infty}^{\infty} (\mu + y)f(\mu + y)\,dy$$

$$= \; \mu \int_{-\infty}^{\infty} f(\mu + y)\,dy \; + \; \int_0^{\infty} y\,(f(\mu + y) - f(\mu - y))\,dy$$

$$= \; (\mu \times 1) + 0 \; = \; \mu.$$

2.4 The Variance of a Random Variable

2.4.1 (a) $E(X) = \left(-2 \times \frac{1}{3}\right) + \left(1 \times \frac{1}{6}\right) + \left(4 \times \frac{1}{3}\right) + \left(6 \times \frac{1}{6}\right) = \frac{11}{6}$.

(b) $Var(X) = \left(\frac{1}{3} \times \left(-2 - \frac{11}{6}\right)^2\right) + \left(\frac{1}{6} \times \left(1 - \frac{11}{6}\right)^2\right)$

$+ \left(\frac{1}{3} \times \left(4 - \frac{11}{6}\right)^2\right) + \left(\frac{1}{6} \times \left(6 - \frac{11}{6}\right)^2\right) = \frac{341}{36}$.

(c) $E(X^2) = \left(\frac{1}{3} \times (-2)^2\right) + \left(\frac{1}{6} \times 1^2\right) + \left(\frac{1}{3} \times 4^2\right) + \left(\frac{1}{6} \times 6^2\right) = \frac{77}{6}$.

$Var(X) = E(X^2) - E(X)^2 = \frac{77}{6} - \left(\frac{11}{6}\right)^2 = \frac{341}{36}$.

2.4.2 $E(X^2) = (0^2 \times 0.08) + (1^2 \times 0.11) + (2^2 \times 0.27) + (3^2 \times 0.33)$

$+ (4^2 \times 0.21) = 7.52$.

Then $E(X) = 2.48$ so $Var(X) = 7.52 - (2.48)^2 = 1.37$ and $\sigma = 1.17$.

2.4.3 $E(X^2) = \left(1^2 \times \frac{2}{5}\right) + \left(2^2 \times \frac{3}{10}\right) + \left(3^2 \times \frac{1}{5}\right) + \left(4^2 \times \frac{1}{10}\right) = 5$.

Then $E(X) = 2$ so $Var(X) = 5 - 2^2 = 1$ and $\sigma = 1$.

2.4.4 See Problem 2.3.9.

$E(X^2) = (0^2 \times 0.168) + (1^2 \times 0.2816) + (3^2 \times 0.1664) + (4^2 \times 0.1024)$

$+ (5^2 \times 0.0512) = 5.6192$.

Then $E(X) = 1.9072$ so $Var(X) = 5.6192 - 1.9072^2 = 1.98$ and $\sigma = 1.41$.
A small variance is generally preferable if the expected winnings are positive.

2.4.5 (a) $E(X^2) = \int_4^6 x^2 \frac{1}{x \ln(1.5)} \, dx = 24.66$.

Then $E(X) = 4.94$ so $Var(X) = 24.66 - 4.94^2 = 0.25$.

(b) $\sigma = \sqrt{0.25} = 0.5$.

(c) Solving $F(x) = 0.25$ gives $x = 4.43$.
Solving $F(x) = 0.75$ gives $x = 5.42$.

(d) Interquartile range $= 5.42 - 4.43 = 0.99$.

2.4.6   (a) $E(X^2) = \int_0^4 x^2 \left(\frac{x}{8}\right) \, dx = 8.$

Then $E(X) = \frac{8}{3}$ so $Var(X) = 8 - \left(\frac{8}{3}\right)^2 = \frac{8}{9}$.

(b) $\sigma = \sqrt{\frac{8}{9}} = 0.94$.

(c) Solving $F(x) = 0.25$ gives $x = 2$.
Solving $F(x) = 0.75$ gives $x = \sqrt{12} = 3.46$. .

(d) Interquartile range $= 3.46 - 2.00 = 1.46$.

2.4.7   (a) $E(X^2) = \int_{0.125}^{0.5} x^2 \, 5.5054(0.5 - (x - 0.25)^2) \, dx = 0.1073.$

Then $E(X) = 0.3095$ so $Var(X) = 0.1073 - 0.3095^2 = 0.0115$.

(b) $\sigma = \sqrt{0.0115} = 0.107$.

(c) Solving $F(x) = 0.25$ gives $x = 0.217$.
Solving $F(x) = 0.75$ gives $x = 0.401$.

(d) Interquartile range $= 0.401 - 0.217 = 0.184$.

2.4.8   (a) $E(X^2) = \int_0^{10} \frac{\theta^2}{e^{10} - 11}(e^{10-\theta} - 1) \, d\theta = 1.9803.$

Then $E(X) = 0.9977$ so $Var(X) = 1.9803 - 0.9977^2 = 0.985$.

(b) $\sigma = \sqrt{0.985} = 0.992$.

(c) Solving $F(\theta) = 0.25$ gives $\theta = 0.288$.
Solving $F(\theta) = 0.75$ gives $\theta = 1.385$.

(d) Interquartile range $= 1.385 - 0.288 = 1.097$.

2.4.9   (a) $E(X^2) = \int_0^{50} \frac{375.3 \, r^2}{(r + 5)^4} \, dr = 18.80.$

Then $E(X) = 2.44$ so $Var(X) = 18.80 - 2.44^2 = 12.8.$

(b) $\sigma = \sqrt{12.8} = 3.58$.

(c) Solving $F(r) = 0.25$ gives $r = 0.50$.
Solving $F(r) = 0.75$ gives $r = 2.93$.

     (d) Interquartile range $= 2.93 - 0.50 = 2.43$.

2.4.10   Adding and subtracting two standard deviations from the mean value gives

$$P(60.4 \leq X \leq 89.6) \geq 0.75.$$

Adding and subtracting three standard deviations from the mean value gives

$$P(53.1 \leq X \leq 96.9) \geq 0.89.$$

2.4.11   The interval $(109.55, 112.05)$ is $(\mu - 2.5c, \mu + 2.5c)$ so Chebyshev's inequality gives

$$P(109.55 \leq X \leq 112.05) \geq 1 - \frac{1}{2.5^2} = 0.84.$$

2.5   Jointly Distributed Random Variables

2.5.1   (a) $P(0.8 \leq X \leq 1, 25 \leq Y \leq 30)$

$$= \int_{x=0.8}^{1} \int_{y=25}^{30} \left(\frac{39}{400} - \frac{17(x-1)^2}{50} - \frac{(y-25)^2}{10,000} \right) dx \, dy \; = \; 0.092.$$

(b) $E(Y) \; = \; \int_{20}^{35} y \left(\frac{83}{1,200} - \frac{(y-25)^2}{10,000} \right) dy \; = \; 27.36.$

$$E(Y^2) \; = \; \int_{20}^{35} y^2 \left(\frac{83}{1,200} - \frac{(y-25)^2}{10,000} \right) dy \; = \; 766.84.$$

$$Var(Y) \; = \; E(Y^2) - E(Y)^2 \; = \; 766.84 - (27.36)^2 \; = \; 18.27.$$

$$\sigma_Y \; = \; \sqrt{18.274} \; = \; 4.27.$$

(c) $E(Y|X = 0.55) \; = \; \int_{20}^{35} y \left(0.073 - \frac{(y-25)^2}{3,922.5} \right) dy \; = \; 27.14.$

$$E(Y^2|X = 0.55) \; = \; \int_{20}^{35} y^2 \left(0.073 - \frac{(y-25)^2}{3,922.5} \right) dy \; = \; 753.74.$$

$$Var(Y|X = 0.55) \; = \; E(Y^2|X = 0.55) - E(Y|X = 0.55)^2$$

$$= \; 753.74 - (27.14)^2 \; = \; 17.16.$$

$$\sigma_{Y|X=0.55} \; = \; \sqrt{17.16} \; = \; 4.14.$$

2.5.2   (a) $p_{1|Y=1} \; = \; P(X = 1|Y = 1) \; = \; \frac{p_{11}}{p_{+1}} \; = \; \frac{0.12}{0.32} \; = \; 0.37500.$

$$p_{2|Y=1} \; = \; P(X = 2|Y = 1) \; = \; \frac{p_{21}}{p_{+1}} \; = \; \frac{0.08}{0.32} \; = \; 0.25000.$$

$$p_{3|Y=1} \; = \; P(X = 3|Y = 1) \; = \; \frac{p_{31}}{p_{+1}} \; = \; \frac{0.07}{0.32} \; = \; 0.21875.$$

$$p_{4|Y=1} \; = \; P(X = 4|Y = 1) \; = \; \frac{p_{41}}{p_{+1}} \; = \; \frac{0.05}{0.32} \; = \; 0.15625.$$

$$E(X|Y=1) \; = \; (1 \times 0.375) + (2 \times 0.25) + (3 \times 0.21875)$$

$$+ \; (4 \times 0.15625) \; = \; 2.15625.$$

$$E(X^2|Y=1) \; = \; = \; (1^2 \times 0.375) + (2^2 \times 0.25) + (3^2 \times 0.21875)$$

$$+ \; (4^2 \times 0.15625) \; = \; 5.84375.$$

$$Var(X|Y=1) \; = \; E(X^2|Y=1) - E(X|Y=1)^2 \; = \; 5.84375 - 2.15625^2$$

$$= \; 1.1943.$$

$$\sigma_{X|Y=1} \; = \; \sqrt{1.1943} \; = \; 1.093.$$

(b) $$p_{1|X=2} \; = \; P(Y=1|X=2) \; = \; \frac{p_{21}}{p_{2+}} \; = \; \frac{0.08}{0.24} \; = \; \frac{8}{24}.$$

$$p_{2|X=2} \; = \; P(Y=2|X=2) \; = \; \frac{p_{22}}{p_{2+}} \; = \; \frac{0.15}{0.24} \; = \; \frac{15}{24}.$$

$$p_{3|X=2} \; = \; P(Y=3|X=2) \; = \; \frac{p_{23}}{p_{2+}} \; = \; \frac{0.01}{0.24} \; = \; \frac{1}{24}.$$

$$E(Y|X=2) \; = \; \left(1 \times \frac{8}{24}\right) + \left(2 \times \frac{15}{24}\right) + \left(3 \times \frac{1}{24}\right) \; = \; \frac{41}{24} \; = \; 1.7083.$$

$$E(Y^2|X=2) \; = \; \left(1^2 \times \frac{8}{24}\right) + \left(2^2 \times \frac{15}{24}\right) + \left(3^2 \times \frac{1}{24}\right) \; = \; \frac{77}{24} \; = \; 3.2083.$$

$$Var(Y|X=2) \; = \; E(Y^2|X=2) - E(Y|X=2)^2$$

$$= \; 3.2083 - 1.7083^2 \; = \; 0.290.$$

$$\sigma_{Y|X=2} \; = \; \sqrt{0.290} \; = \; 0.538.$$

2.5.3   (a) $$\int_{x=-2}^{3} \int_{y=4}^{6} A(x-3)y \; dx \; dy \; = \; 1 \quad \Rightarrow \quad A = -\frac{1}{125}.$$

(b) $$P(0 \le X \le 1, 4 \le Y \le 5) \; = \; \int_{x=0}^{1} \int_{y=4}^{5} \frac{(3-x)y}{125} \; dx \; dy \; = \; \frac{9}{100}.$$

(c) $\qquad f_X(x) = \int_4^6 \frac{(3-x)y}{125} \, dy = \frac{2(3-x)}{25} \quad$ for $\quad -2 \le x \le 3.$

$\qquad\qquad f_Y(y) = \int_{-2}^3 \frac{(3-x)y}{125} \, dx = \frac{y}{10} \quad$ for $\quad 4 \le x \le 6.$

(d) They are independent since $f_X(x) \times f_Y(y) = f(x,y)$ and the ranges of the random variables are not connected with each other.

(e) $f_{X|Y=5}(x)$ is equal to $f_X(x)$ since the random variables are independent.

2.5.4 (a)

X\Y	0	1	2	3	p_{i+}
0	1/16	1/16	0	0	2/16
1	1/16	3/16	2/16	0	6/16
2	0	2/16	3/16	1/16	6/16
3	0	0	1/16	1/16	2/16
p_{+j}	2/16	6/16	6/16	2/16	

(b) See table above.

(c) Notice that $p_{00} = \frac{1}{16}$ while $p_{0+} \times p_{+0} = \frac{2}{16} \times \frac{2}{16} = \frac{1}{4} \ne p_{00}$ so the random variables X and Y are not independent.

(d) $\qquad E(X) = \left(0 \times \frac{2}{16}\right) + \left(1 \times \frac{6}{16}\right) + \left(2 \times \frac{6}{16}\right) + \left(3 \times \frac{2}{16}\right) = \frac{3}{2}.$

$\qquad E(X^2) = \left(0^2 \times \frac{2}{16}\right) + \left(1^2 \times \frac{6}{16}\right) + \left(2^2 \times \frac{6}{16}\right) + \left(3^2 \times \frac{2}{16}\right) = 3.$

$\qquad Var(X) = E(X^2) - E(X)^2 = 3 - \left(\frac{3}{2}\right)^2 = \frac{3}{4}.$

The random variable Y has the same mean and variance.

(e) $\qquad E(XY) = \left(1 \times 1 \times \frac{3}{16}\right) + \left(1 \times 2 \times \frac{2}{16}\right) + \left(2 \times 1 \times \frac{2}{16}\right)$

$\qquad\qquad + \left(2 \times 2 \times \frac{3}{16}\right) + \left(2 \times 3 \times \frac{1}{16}\right) + \left(3 \times 2 \times \frac{1}{16}\right) + \left(3 \times 3 \times \frac{1}{16}\right)$

$\qquad\qquad = \frac{44}{16}.$

$\qquad Cov(X,Y) = E(XY) - (E(X) \times E(Y)) = \frac{44}{16} - \left(\frac{3}{2} \times \frac{3}{2}\right) = \frac{1}{2}.$

(f) $\qquad P(X = 0|Y = 1) = \frac{p_{01}}{p_{+1}} = \frac{\left(\frac{1}{16}\right)}{\left(\frac{6}{16}\right)} = \frac{1}{6}.$

$\qquad\qquad P(X = 1|Y = 1) = \frac{p_{11}}{p_{+1}} = \frac{\left(\frac{3}{16}\right)}{\left(\frac{6}{16}\right)} = \frac{1}{2}.$

$$P(X = 2 | Y = 1) = \frac{p_{21}}{p_{+1}} = \frac{\left(\frac{2}{16}\right)}{\left(\frac{6}{16}\right)} = \frac{1}{3}.$$

$$P(X = 3 | Y = 1) = \frac{p_{31}}{p_{+1}} = \frac{0}{\left(\frac{6}{16}\right)} = 0.$$

$$E(X | Y = 1) = \left(0 \times \frac{1}{6}\right) + \left(1 \times \frac{1}{2}\right) + \left(2 \times \frac{1}{3}\right) = \frac{7}{6}.$$

$$E(X^2 | Y = 1) = \left(0^2 \times \frac{1}{6}\right) + \left(1^2 \times \frac{1}{2}\right) + \left(2^2 \times \frac{1}{3}\right) = \frac{11}{6}$$

$$V(X | Y = 1) = E(X^2 | Y = 1) - E(X | Y = 1)^2 = \frac{11}{6} - \left(\frac{7}{6}\right)^2 = \frac{17}{36}.$$

2.5.5 (a) $\displaystyle\int_{x=1}^{2} \int_{y=0}^{3} A(e^{x+y} + e^{2x-y}) \, dx \, dy = 1 \quad \Rightarrow \quad A = 0.00896.$

(b) $P(1.5 \leq X \leq 2, 1 \leq Y \leq 2) =$

$$\int_{x=1.5}^{2} \int_{y=1}^{2} 0.00896 \, (e^{x+y} + e^{2x-y}) \, dx \, dy = 0.158.$$

(c) $\displaystyle f_X(x) = \int_0^3 0.00896 \, (e^{x+y} + e^{2x-y}) \, dy$

$$= 0.00896 \, (e^{x+3} - e^{2x-3} - e^x + e^{2x}) \quad \text{for} \quad 1 \leq x \leq 2.$$

$$f_Y(y) = \int_1^2 0.00896 \, (e^{x+y} + e^{2x-y}) \, dx$$

$$= 0.00896 \, (e^{2+y} + 0.5e^{4-y} - e^{1+y} - 0.5e^{2-y}) \quad \text{for} \quad 0 \leq y \leq 3.$$

(d) No, $f_X(x) \times f_Y(y) \neq f(x, y).$

(e) $\displaystyle f_{X|Y=0}(x) = \frac{f(x, 0)}{f_Y(0)} = \frac{e^x + e^{2x}}{28.28}.$

2.5.6 (a)

X\Y	0	1	2	p_{i+}
0	25/102	26/102	6/102	57/102
1	26/102	13/102	0	39/102
2	6/102	0	0	6/102
p_{+j}	57/102	39/102	6/102	

(b) See table above.

(c) No they are not independent. For example $p_{22} \neq p_{2+} \times p_{+2}$.

(d)
$$E(X) = \left(0 \times \frac{57}{102}\right) + \left(1 \times \frac{39}{102}\right) + \left(2 \times \frac{6}{102}\right) = \frac{1}{2}.$$

$$E(X^2) = \left(0^2 \times \frac{57}{102}\right) + \left(1^2 \times \frac{39}{102}\right) + \left(2^2 \times \frac{6}{102}\right) = \frac{21}{34}.$$

$$Var(X) = E(X^2) - E(X)^2 = \frac{21}{34} - \left(\frac{1}{2}\right)^2 = \frac{25}{68}.$$

The random variable Y has the same mean and variance.

(e)
$$E(XY) = 1 \times 1 \times p_{11} = \frac{13}{102}.$$

$$Cov(X,Y) = E(XY) - (E(X) \times E(Y)) = \frac{13}{102} - \left(\frac{1}{2} \times \frac{1}{2}\right) = -\frac{25}{204}.$$

(f)
$$Corr(X,Y) = \frac{Cov(X,Y)}{\sqrt{Var(X)Var(Y)}} = -\frac{1}{3}.$$

(g)
$$P(Y=0|X=0) = \frac{p_{00}}{p_{0+}} = \frac{25}{57}.$$

$$P(Y=1|X=0) = \frac{p_{01}}{p_{0+}} = \frac{26}{57}.$$

$$P(Y=2|X=0) = \frac{p_{02}}{p_{0+}} = \frac{6}{57}.$$

$$P(Y=0|X=1) = \frac{p_{10}}{p_{1+}} = \frac{2}{3}.$$

$$P(Y=1|X=1) = \frac{p_{11}}{p_{1+}} = \frac{1}{3}.$$

$$P(Y=2|X=1) = \frac{p_{12}}{p_{1+}} = 0.$$

2.5.7   (a)

X\Y	0	1	2	p_{i+}
0	4/16	4/16	1/16	9/16
1	4/16	2/16	0	6/16
2	1/16	0	0	1/16
p_{+j}	9/16	6/16	1/16	

(b) See table above.

(c) No they are not independent. For example $p_{22} \neq p_{2+} \times p_{+2}$.

(d)
$$E(X) = \left(0 \times \frac{9}{16}\right) + \left(1 \times \frac{6}{16}\right) + \left(2 \times \frac{1}{16}\right) = \frac{1}{2}.$$

$$E(X^2) = \left(0^2 \times \frac{9}{16}\right) + \left(1^2 \times \frac{6}{16}\right) + \left(2^2 \times \frac{1}{16}\right) = \frac{5}{8}.$$

$$Var(X) = E(X^2) - E(X)^2 = \frac{5}{8} - \left(\frac{1}{2}\right)^2 = \frac{3}{8} = 0.3676.$$

The random variable Y has the same mean and variance.

(e)
$$E(XY) = 1 \times 1 \times p_{11} = \frac{1}{8}.$$

$$Cov(X,Y) = E(XY) - (E(X) \times E(Y)) = \frac{1}{8} - \left(\frac{1}{2} \times \frac{1}{2}\right) = -\frac{1}{8}.$$

(f)
$$Corr(X,Y) = \frac{Cov(X,Y)}{\sqrt{Var(X)Var(Y)}} = -\frac{1}{3}.$$

(g)
$$P(Y = 0 | X = 0) = \frac{p_{00}}{p_{0+}} = \frac{4}{9}.$$

$$P(Y = 1 | X = 0) = \frac{p_{01}}{p_{0+}} = \frac{4}{9}.$$

$$P(Y = 2 | X = 0) = \frac{p_{02}}{p_{0+}} = \frac{1}{9}.$$

$$P(Y = 0 | X = 1) = \frac{p_{10}}{p_{1+}} = \frac{2}{3}.$$

$$P(Y = 1 | X = 1) = \frac{p_{11}}{p_{1+}} = \frac{1}{3}.$$

$$P(Y = 2 | X = 1) = \frac{p_{12}}{p_{1+}} = 0.$$

2.5.8   (a)
$$\int_{x=0}^{5} \int_{y=0}^{5} A\,(20 - x - 2y)\, dx\, dy = 1 \quad \Rightarrow \quad A = 0.0032.$$

(b) $P(1 \leq X \leq 2, 2 \leq Y \leq 3) = \int_{x=1}^{2} \int_{y=2}^{3} 0.0032\,(20 - x - 2y)\,dx\,dy$

           $= 0.0432.$

(c) $f_X(x) = \int_{y=0}^{5} 0.0032\,(20 - x - 2y)\,dy = 0.016\,(15 - x)$

           for $0 \leq x \leq 5.$

           $f_Y(y) = \int_{x=0}^{5} 0.0032\,(20 - x - 2y)\,dx = 0.008\,(35 - 4y)$

           for $0 \leq y \leq 5.$

(d) No they are not independent since $f(x,y) \neq f_X(x)f_Y(y).$

(e) $E(X) = \int_{0}^{5} x\,0.016\,(15 - x)\,dx = \dfrac{7}{3}.$

           $E(X^2) = \int_{0}^{5} x^2\,0.016\,(15 - x)\,dx = \dfrac{15}{2}.$

           $Var(X) = E(X^2) - E(X)^2 = \dfrac{15}{2} - \left(\dfrac{7}{3}\right)^2 = \dfrac{37}{18}.$

(f) $E(Y) = \int_{0}^{5} y\,0.008\,(35 - 4y)\,dy = \dfrac{13}{6}.$

           $E(Y^2) = \int_{0}^{5} y^2\,0.008\,(35 - 4y)\,dy = \dfrac{20}{3}.$

           $Var(Y) = E(Y^2) - E(Y)^2 = \dfrac{20}{3} - \left(\dfrac{13}{6}\right)^2 = \dfrac{71}{36}.$

(g) $f_{Y|X=3}(y) = \dfrac{f(3,y)}{f_X(3)} = \dfrac{17 - 2y}{60}$ for $0 \leq y \leq 5.$

(h) $E(XY) = \int_{x=0}^{5} \int_{y=0}^{5} 0.0032\,xy\,(20 - x - 2y)\,dx\,dy = 5.$

           $Cov(X,Y) = E(XY) - (E(X) \times (EY)) = 5 - \left(\dfrac{7}{3} \times \dfrac{13}{6}\right) = -\dfrac{1}{18}.$

(i) $Corr(X,Y) = \dfrac{Cov(X,Y)}{\sqrt{Var(X)Var(Y)}} = -0.0276$

2.5.9   (a) P(same score) = P(X=1,Y=1) + P(X=2,Y=2) + P(X=3,Y=3)
            + P(X=4,Y=4) = 0.80.

    (b) P(X < Y) = P(X=1,Y=2) + P(X=1,Y=3) + P(X=1,Y=4) + P(X=2,Y=3)
            + P(X=2,Y=4) + P(X=3,Y=4) = 0.07.

    (c)

x_i	1	2	3	4
p_{i+}	0.12	0.20	0.30	0.38

$$E(X) = (1 \times 0.12) + (2 \times 0.20) + (3 \times 0.30) + (4 \times 0.38) = 2.94.$$

$$E(X^2) = (1^2 \times 0.12) + (2^2 \times 0.20) + (3^2 \times 0.30) + (4^2 \times 0.38)$$

$$= 9.70.$$

$$Var(X) = E(X^2) - E(X)^2 = 9.70 - (2.94)^2 = 1.0564.$$

    (d)

y_j	1	2	3	4
p_{+j}	0.14	0.21	0.30	0.35

$$E(Y) = (1 \times 0.14) + (2 \times 0.21) + (3 \times 0.30) + (4 \times 0.35) = 2.86.$$

$$E(Y^2) = (1^2 \times 0.14) + (2^2 \times 0.21) + (3^2 \times 0.30) + (4^2 \times 0.35)$$

$$= 9.28.$$

$$Var(Y) = E(Y^2) - E(Y)^2 = 9.28 - (2.86)^2 = 1.1004.$$

    (e) The scores are not independent. For example $p_{11} \neq p_{1+} \times p_{+1}$. The scores would not be expected to be independent since they apply to the two inspector's assessments of the same building. If they were independent it would suggest that one of the inspectors is randomly assigning a safety score without paying any attention to the actual state of the building.

    (f)

$$P(Y = 1 | X = 3) = \frac{p_{31}}{p_{3+}} = \frac{1}{30}.$$

$$P(Y = 2 | X = 3) = \frac{p_{32}}{p_{3+}} = \frac{3}{30}.$$

$$P(Y = 3 | X = 3) = \frac{p_{33}}{p_{3+}} = \frac{24}{30}.$$

$$P(Y = 4 | X = 3) = \frac{p_{34}}{p_{3+}} = \frac{2}{30}.$$

    (g)

$$E(XY) = \sum_{i=1}^{4}\sum_{j=1}^{4} i\, j\, p_{ij} = 9.29.$$

$$Cov(X,Y) = E(XY) - (E(X) \times E(Y)) = 9.29 - (2.94 \times 2.86) = 0.8816.$$

(h) $Corr(X, Y) = \dfrac{Cov(X, Y)}{\sqrt{VarX \; VarY}} = \dfrac{0.8816}{\sqrt{1.0564 \times 1.1004}} = 0.82.$

A high positive correlation indicates that the inspectors are consistent. The closer the correlation is to one the more consistent are the inspectors.

2.5.10   (a) $\displaystyle\int_{x=0}^{2} \int_{y=0}^{2} \int_{z=0}^{2} \dfrac{3xyz^2}{32} \; dx \; dy \; dz = 1.$

(b) $\displaystyle\int_{x=0}^{1} \int_{y=0.5}^{1.5} \int_{z=1}^{2} \dfrac{3xyz^2}{32} \; dx \; dy \; dz = \dfrac{7}{64}.$

(c) $f_X(x) = \displaystyle\int_{x=0}^{2} \int_{y=0}^{2} \dfrac{3xyz^2}{32} \; dy \; dz = \dfrac{x}{2} \quad \text{for} \quad 0 \leq 2 \leq x.$

2.6 Combinations and Functions of Random Variables

2.6.1   (a) $E(3X + 7) = 3E(X) + 7 = 13.$
        $Var(3X + 7) = 3^2 Var(X) = 36.$

  (b) $E(5X - 9) = 5E(X) - 9 = 1.$
        $Var(5X - 9) = 5^2 Var(X) = 100.$

  (c) $E(2X + 6Y) = 2E(X) + 6E(Y) = -14.$
        $Var(2X + 6Y) = 2^2 Var(X) + 6^2 Var(Y) = 88.$

  (d) $E(4X - 3Y) = 4E(X) - 3E(Y) = 17.$
        $Var(4X - 3Y) = 4^2 Var(X) + 3^2 Var(Y) = 82.$

  (e) $E(5X - 9Z + 8) = 5E(X) - 9E(Z) + 8 = -54.$
        $Var(5X - 9Z + 8) = 5^2 Var(X) + 9^2 Var(Z) = 667.$

  (f) $E(-3Y - Z - 5) = -3E(Y) - E(Z) - 5 = -4.$
        $Var(-3Y - Z - 5) = (-3)^2 Var(Y) + (-1)^2 Var(Z) = 25.$

  (g) $E(X + 2Y + 3Z) = E(X) + 2E(Y) + 3E(Z) = 20.$
        $Var(X + 2Y + 3Z) = Var(X) + 2^2 Var(Y) + 3^2 Var(Z) = 75.$

  (h) $E(6X + 2Y - Z + 16) = 6E(X) + 2E(Y) - E(Z) + 16 = 14.$
        $Var(6X + 2Y - Z + 16) = 6^2 Var(X) + 2^2 Var(Y) + (-1)^2 Var(Z) = 159.$

2.6.2 $E(aX + b) = \int (ax + b) \, f(x) \, dx = a \int x \, f(x) \, dx + b \int f(x) \, dx$

$$= aE(X) + b.$$

$Var(aX + b) = E((aX + b - E(aX + b))^2) = E((aX - aE(X))^2)$
$= a^2 E((X - E(X))^2) = a^2 Var(X).$

2.6.3 $E(Y) = 3E(X_1) = 3\mu.$
     $Var(Y) = 3^2 Var(X_1) = 9\sigma^2.$

$E(Z) = E(X_1) + E(X_2) + E(X_3) = 3\mu$
$Var(Z) = Var(X_1) + Var(X_2) + Var(X_3) = 3\sigma^2.$

The random variables Y and Z have the same mean but Z has a smaller variance than Y.

2.6.4 $length = A_1 + A_2 + B.$
$E(length) = E(A_1) + E(A_2) + E(B) = 37 + 37 + 24 = 98.$
$Var(length) = Var(A_1) + Var(A_2) + V(B) = 0.7^2 + 0.7^2 + 0.3^2 = 1.07.$

2.6.5   Let the random variable X_i be the winnings from the ith game. Then

$$E(X_i) = \left(10 \times \frac{1}{8}\right) + \left((-1) \times \frac{7}{8}\right) = \frac{3}{8}$$

and

$$E(X_i^2) = \left(10^2 \times \frac{1}{8}\right) + \left((-1)^2 \times \frac{7}{8}\right) = \frac{107}{8}$$

so that

$$Var(X_i) = E(X_i^2) - (E(X_i))^2 = \frac{847}{64}.$$

The total winnings from 50 (independent) games is $Y = X_1 + \ldots + X_{50}$ and

$$E(Y) = E(X_1) + \ldots + E(X_{50}) = 50 \times \frac{3}{8} = \frac{75}{4} = \$18.75$$

with

$$Var(Y) = Var(X_1) + \ldots + Var(X_{50}) = 50 \times \frac{847}{64} = 661.72$$

so that $\sigma_Y = \sqrt{661.72} = \$25.72$.

2.6.6   (a) $E(average\ weight) = 1.12$ kg.
$Var(average\ weight) = \frac{0.03^2}{25} = 3.6 \times 10^{-5}$ and the standard deviation is
$\frac{0.03}{\sqrt{25}} = 0.0012$ kg.

(b) It is required that $\frac{0.03}{\sqrt{n}} \le 0.005$ which is satisfied for $n \ge 36$.

2.6.7   Let the random variable X_i be equal to 1 if an ace is drawn on the ith drawing (with probability $\frac{1}{13}$) and equal to 0 if an ace is not drawn on the ith drawing (with probability $\frac{12}{13}$). Then the total number of aces drawn is $Y = X_1 + \ldots + X_{10}$.

Notice that $E(X_i) = \frac{1}{13}$. Regardless of whether the drawing is performed with or without replacement
$E(Y) = E(X_1) + \ldots + E(X_{10}) = \frac{10}{13}$.
Notice also that $E(X_i^2) = \frac{1}{13}$ so that

$$Var(X_i) = \frac{1}{13} - \left(\frac{1}{13}\right)^2 = \frac{12}{169}.$$

If the drawings are made *with* replacement then the random variables X_i are independent so that
$Var(Y) = Var(X_1) + \ldots + Var(X_{10}) = \frac{120}{169}$.
If the drawings are made *without* replacement then the random variables X_i are not independent.

2.6.8 $F_X(x) = P(X \le x) = x^2$ for $0 \le x \le 1$.

(a) $F_Y(y) = P(Y \le y) = P(X^3 \le y) = P(X \le y^{1/3}) = F_X(y^{1/3}) = y^{2/3}$ and so $f_Y(y) = \frac{2}{3} y^{-1/3}$ for $0 \le y \le 1$ and

$$E(y) = \int_0^1 y \, f_Y(y) \, dy = 0.4.$$

(b) $F_Y(y) = P(Y \le y) = P(\sqrt{X} \le y) = P(X \le y^2) = F_X(y^2) = y^4$ and so $f_Y(y) = 4y^3$ for $0 \le y \le 1$ and

$$E(y) = \int_0^1 y \, f_Y(y) \, dy = 0.8.$$

(c) $F_Y(y) = P(Y \le y) = P(\frac{1}{1+X} \le y) = P(X \ge \frac{1}{y} - 1) = 1 - F_X(\frac{1}{y} - 1)$ $= \frac{2}{y} - \frac{1}{y^2}$ and so

$$f_Y(y) = -\frac{2}{y^2} + \frac{2}{y^3} \quad \text{for} \quad \frac{1}{2} \le y \le 1$$

and

$$E(y) = \int_{0.5}^1 y \, f_Y(y) \, dy = 0.614.$$

(d) $F_Y(y) = P(Y \le y) = P(2^X \le y) = P\left(X \le \frac{\ln(y)}{\ln(2)}\right) = F_X\left(\frac{\ln(y)}{\ln(2)}\right) = \left(\frac{\ln(y)}{\ln(2)}\right)^2$ and so

$$f_Y(y) = \frac{2 \ln(y)}{y \, (\ln(2))^2} \quad \text{for} \quad 1 \le y \le 2$$

and

$$E(y) = \int_1^2 y \, f_Y(y) \, dy = 1.61.$$

2.6.9 (a) $\int_0^2 A(1 - (r-1)^2) \, dr = 1 \quad \Rightarrow \quad A = \frac{3}{4}.$

This gives

$$F_R(r) = \frac{3r^2}{4} - \frac{r^3}{4} \quad \text{for} \quad 0 \le r \le 2.$$

(b) $V = \frac{4}{3}\pi r^3$.

$$F_V(v) = P(V \le v) = P\left(\frac{4}{3}\pi r^3 \le v\right) = F_R\left(\left(\frac{3v}{4\pi}\right)^{1/3}\right).$$

This gives

$$f_V(v) = \frac{1}{2}(\frac{3}{4\pi})^{2/3}v^{-1/3} - \frac{3}{16\pi} \quad \text{for} \quad 0 \le v \le \frac{32\pi}{3}.$$

(c) $\qquad E(V) = \int_0^{\frac{32\pi}{3}} v\, f_V(v)\, dv = \frac{32\pi}{15}.$

2.6.10   (a) $\quad \int_0^L Ax(L-x)\, dx = 1 \quad \Rightarrow \quad A = \frac{6}{L^3}.$

This gives

$$F_X(x) = \frac{x^2(3L - 2x)}{L^3} \quad \text{for} \quad 0 \le x \le L.$$

(b) The random variable corresponding to the difference in the lengths of the two pieces of rod is $W = |L - 2X|$.

$$F_W(w) = P\left(\frac{L}{2} - \frac{w}{2} \le X \le \frac{L}{2} + \frac{w}{2}\right) = F_X\left(\frac{L}{2} + \frac{w}{2}\right) - F_X\left(\frac{L}{2} - \frac{w}{2}\right)$$

$$= \frac{w(3L^2 - w^2)}{2L^3}.$$

This gives

$$f_W(w) = \frac{3(L^2 - w^2)}{2L^3} \quad \text{for} \quad 0 \le w \le L.$$

(c) $\qquad E(W) = \int_0^L w\, f_W(w)\, dw = \frac{3}{8}L.$

2.6.11   (a) The return has an expectation of $100, a standard deviation of $20 and a variance of 400.

(b) The return has an expectation of $100, a standard deviation of $30 and a variance of 900.

(c) The return from fund A has an expectation of $50, a standard deviation of $10 and a variance of 100. The return from fund B has an expectation of $50, a standard deviation of $15 and a variance of 225. The total return therefore has an expectation of $100 and and a variance of 325.

(d) The return from fund A has an expectation of $0.1x$, a standard deviation of $0.02x$ and a variance of $0.0004x^2$. The return from fund B has an expectation of $0.1(1000 - x)$, a standard deviation of $0.03(1000 - x)$ and a variance of

$0.0009(1000 - x)^2$. The total return therefore has an expectation of $100 and and a variance of $0.0004x^2 + 0.0009(1000 - x)^2$. This variance is minimized by taking $x = \$692$ and the minimum value of the variance is 276.9 corresponding to a standard deviation of $16.64.

This problem illustrates that the variability of the return on an investment can be reduced by diversifying the investment, that is spreading it over several funds.

2.7 Supplementary Problems

2.7.1 (a)

x_i	2	3	4	5	6
p_i	$\frac{1}{15}$	$\frac{2}{15}$	$\frac{3}{15}$	$\frac{4}{15}$	$\frac{5}{15}$

(b) $$E(X) = \left(2 \times \frac{2}{30}\right) + \left(3 \times \frac{4}{30}\right) + \left(4 \times \frac{6}{30}\right) + \left(5 \times \frac{8}{30}\right) + \left(6 \times \frac{10}{30}\right)$$
$$= \frac{14}{3}.$$

2.7.2 (a)

x_i	0	1	2	3	4	5	6
$F(x_i)$	0.21	0.60	0.78	0.94	0.97	0.99	1.00

(b) $$E(X) = (0 \times 0.21) + (1 \times 0.39) + (2 \times 0.18) + (3 \times 0.16)$$
$$+ (4 \times 0.03) + (5 \times 0.02) + (6 \times 0.01) = 1.51.$$

(c) $$E(X^2) = (0^2 \times 0.21) + (1^2 \times 0.39) + (2^2 \times 0.18) + (3^2 \times 0.16)$$
$$+ (4^2 \times 0.03) + (5^2 \times 0.02) + (6^2 \times 0.01) = 3.89.$$
$$Var(X) = 3.89 - (1.51)^2 = 1.61.$$

(d) The expectation is $1.51 \times 60 = 90.6$ and the variance is $1.61 \times 60 = 96.6$.

2.7.3 (a)

x_i	2	3	4	5
p_i	$\frac{2}{30}$	$\frac{13}{30}$	$\frac{13}{30}$	$\frac{2}{30}$

(b) $$E(X) = \left(2 \times \frac{2}{30}\right) + \left(3 \times \frac{13}{30}\right) + \left(4 \times \frac{13}{30}\right) + \left(5 \times \frac{2}{30}\right) = \frac{7}{2}.$$
$$E(X^2) = \left(2^2 \times \frac{2}{30}\right) + \left(3^2 \times \frac{13}{30}\right) + \left(4^2 \times \frac{13}{30}\right) + \left(5^2 \times \frac{2}{30}\right) = \frac{383}{30}.$$
$$Var(X) = E(X^2) - E(X)^2 = \frac{383}{30} - \left(\frac{7}{2}\right)^2 = \frac{31}{60}.$$

(c)

x_i	2	3	4	5
p_i	$\frac{2}{10}$	$\frac{3}{10}$	$\frac{3}{10}$	$\frac{2}{10}$

$$E(X) = \left(2 \times \frac{2}{10}\right) + \left(3 \times \frac{3}{10}\right) + \left(4 \times \frac{3}{10}\right) + \left(5 \times \frac{2}{10}\right) = \frac{7}{2}.$$

$$E(X^2) = \left(2^2 \times \frac{2}{10}\right) + \left(3^2 \times \frac{3}{10}\right) + \left(4^2 \times \frac{3}{10}\right) + \left(5^2 \times \frac{2}{10}\right) = \frac{133}{10}.$$

$$Var(X) = E(X^2) - E(X)^2 = \frac{133}{10} - \left(\frac{7}{2}\right)^2 = \frac{21}{20}.$$

2.7.4   Let X_i be the value of the ith card dealt. Then

$$E(X_i) = \left(2 \times \frac{1}{13}\right) + \left(3 \times \frac{1}{13}\right) + \left(4 \times \frac{1}{13}\right) + \left(5 \times \frac{1}{13}\right) + \left(6 \times \frac{1}{13}\right)$$

$$+ \left(7 \times \frac{1}{13}\right) + \left(8 \times \frac{1}{13}\right) + \left(9 \times \frac{1}{13}\right) + \left(10 \times \frac{1}{13}\right) + \left(15 \times \frac{4}{13}\right) = \frac{114}{13}.$$

The total score of the hand is $Y = X_1 + \ldots + X_{13}$ which has an expectation

$$E(Y) = E(X_1) + \ldots + E(X_{13}) = 13 \times \frac{114}{13} = 114.$$

2.7.5   (a) $\int_1^{11} A \left(\frac{3}{2}\right)^x dx = 1 \quad \Rightarrow \quad A = \frac{\ln(1.5)}{1.5^{11} - 1.5} = \frac{1}{209.6}.$

(b) $F(x) = \int_1^x \frac{1}{209.6} \left(\frac{3}{2}\right)^y dy = 0.01177 \left(\frac{3}{2}\right)^x - 0.01765$
for $1 \le x \le 11$.

(c) Solving $F(x) = 0.5$ gives $x = 9.332$.

(d) Solving $F(x) = 0.25$ gives $x = 7.706$.
Solving $F(x) = 0.75$ gives $x = 10.305$.
The interquartile range is $10.30 - 7.71 = 2.599$.

2.7.6   (a) $f_X(x) = \int_1^2 4x(2 - y)\, dy = 2x \quad \text{for} \quad 0 \le x \le 1.$

(b) $f_Y(y) = \int_0^1 4x(2 - y)\, dx = 2(2 - y) \quad \text{for} \quad 1 \le y \le 2.$

Since $f(x,y) = f_X(x) \times f_Y(y)$ the random variables are independent.

(c) $Cov(X, Y) = 0$ since the random variables are independent.

(d) $f_{X|Y=1.5}(x) = f_X(x)$ since the random variables are independent.

2.7.7   (a) $\int_5^{10} A\left(x + \frac{2}{x}\right) dx = 1 \quad \Rightarrow \quad A = 0.02572.$

(b) $F(x) = \int_5^x 0.02572\left(y + \frac{2}{y}\right) dy = 0.0129x^2 + 0.0514\ln(x) - 0.404$
for $5 \leq x \leq 10$.

(c) $E(X) = \int_5^{10} 0.02572\, x\left(x + \frac{2}{x}\right) = 7.759.$

(d) $E(X^2) = \int_5^{10} 0.02572\, x^2\left(x + \frac{2}{x}\right) = 62.21.$

$Var(X) = E(X^2) - E(X)^2 = 62.21 - 7.759^2 = 2.01.$

(e) Solving $F(x) = 0.5$ gives $x = 7.88$.

(f) Solving $F(x) = 0.25$ gives $x = 6.58$.
Solving $F(x) = 0.75$ gives $x = 9.00$.
The interquartile range is $9.00 - 6.58 = 2.42$.

(g) The expectation is $E(X) = 7.759$.
The variance is $Var(X)/10 = 0.0201$.

2.7.8 $Var(a_1 X_1 + a_2 X_2 + \ldots + a_n X_n + b)$

$= Var(a_1 X_1) + \ldots + Var(a_n X_n) + Var(b)$

$= a_1^2 Var(X_1) + \ldots + a_n^2 Var(X_n) + 0.$

2.7.9 $Y = \frac{5}{3}X - 25.$

2.7.10   Notice that $E(Y) = aE(X) + b$ and $Var(Y) = a^2 Var(X)$. Also

$Cov(X,Y) = E(\,(X - E(X))\,(Y - E(Y))\,) = E(\,(X - E(X))\,a(X - E(X))\,)$

$= aVar(X).$

Therefore

$$Corr(X,Y) = \frac{Cov(X,Y)}{\sqrt{Var(X)Var(Y)}} = \frac{aVar(X)}{\sqrt{Var(X)a^2Var(X)}} = \frac{a}{|a|}$$

which is 1 if $a > 0$ and -1 if $a < 0$.

2.7.11   The expected amount of a claim is

$$E(X) = \int_0^{1,800} x\,\frac{x(1,800 - x)}{972,000,000}\,dx = \$900.$$

Consequently the expected profit from each customer is $\$100 - \$5 - (0.1 \times \$900) = \$5$. The expected profit from 10,000 customers is $10,000 \times \$5 = \$50,000$.

The profits may or may not be independent depending on the type of insurance and pool of customers. If large natural disasters affect the customers then the claims would not be independent.

Chapter 3

Discrete Probability Distributions

3.1 The Binomial Distribution

3.1.1 (a) $P(X = 3) = \binom{10}{3} \times 0.12^3 \times 0.88^7 = 0.0847.$

(b) $P(X = 6) = \binom{10}{6} \times 0.12^6 \times 0.88^4 = 0.0004.$

(c) $P(X \leq 2) = P(X = 0) + P(X = 1) + P(X = 2) = 0.2785 + 0.3798 + 0.2330$
$= 0.8913.$

(d) $P(X \geq 7) = P(X = 7) + P(X = 8) + P(X = 9) + P(X = 10) = 3.085 \times 10^{-5}.$

(e) $E(X) = 10 \times 0.12 = 1.2.$

(f) $Var(X) = 10 \times 0.12 \times 0.88 = 1.056.$

3.1.2 (a) $P(X = 4) = \binom{7}{4} \times 0.8^4 \times 0.2^3 = 0.1147.$

(b) $P(X \neq 2) = 1 - P(X = 2) = 1 - \binom{7}{2} \times 0.8^2 \times 0.2^5 = 0.9957.$

(c) $P(X \leq 3) = P(X = 0) + P(X = 1) + P(X = 2) + P(X = 3) = 0.0334.$

(d) $P(X \geq 6) = P(X = 6) + P(X = 7) = 0.5767.$

(e) $E(X) = 7 \times 0.8 = 5.6.$

(f) $Var(X) = 7 \times 0.8 \times 0.2 = 1.12.$

3.1.3 $X \sim B(6, 0.5)$

x_i	0	1	2	3	4	5	6
p_i	0.0156	0.0937	0.2344	0.3125	0.2344	0.0937	0.0156

$E(X) = 6 \times 0.5 = 3$ and $Var(X) = 6 \times 0.5 \times 0.5 = 1.5$ with $\sigma = \sqrt{1.5} = 1.22.$

$X \sim B(6, 0.7)$

x_i	0	1	2	3	4	5	6
p_i	0.0007	0.0102	0.0595	0.1852	0.3241	0.3025	0.1176

$E(X) = 6 \times 0.7 = 4.2$ and $Var(X) = 6 \times 0.7 \times 0.3 = 1.26$ with $\sigma = \sqrt{1.5} = 1.12$.

3.1.4 $X \sim B(9, 0.09)$.

  (a) $P(X = 2) = 0.1507$.
  (b) $P(X \geq 2) = 1 - P(X = 0) - P(X = 1) = 0.1912$.

$E(X) = 9 \times 0.09 = 0.81$.

3.1.5   (a) $P(B(8, 0.5) = 5) = 0.2187$.
        (b) $P(B(8, 1/6) = 1) = 0.3721$.
        (c) $P(B(8, 1/6) = 0) = 0.2326$.
        (d) $P(B(8, 2/3) \geq 6) = 0.4682$.

3.1.6 $P(B(10, 0.2) \geq 7) = 0.0009$.
        $P(B(10, 0.5) \geq 7) = 0.1719$.

3.1.7   Let the random variable X be the number of employees taking sick leave. Then $X \sim B(180, 0.35)$. The *proportion* of the workforce who need to take sick leave is

$$Y = \frac{X}{180}.$$

Then

$$E(Y) = \frac{E(X)}{180} = \frac{180 \times 0.35}{180} = 0.35$$

and

$$Var(Y) = \frac{Var(X)}{180^2} = \frac{180 \times 0.35 \times 0.65}{180^2} = 0.0013.$$

In general the variance is

$$Var(Y) = \frac{Var(X)}{180^2} = \frac{180 \times p \times (1-p)}{180^2} = \frac{p \times (1-p)}{180}$$

which is maximized when $p = 0.5$.

3.1.8   The random variable Y can be considered to be the number of successes out of $n_1 + n_2$ trials.

3.2 The Geometric and Negative Binomial Distributions

3.2.1　(a)　$P(X = 4) = (1 - 0.7)^3 \times 0.7 = 0.0189.$

　　　(b)　$P(X = 1) = (1 - 0.7)^0 \times 0.7 = 0.7.$

　　　(c)　$P(X \le 5) = 1 - (1 - 0.7)^5 = 0.9976.$

　　　(d)　$P(X \ge 8) = 1 - P(X \le 7) = (1 - 0.7)^7 = 0.0002.$

3.2.2　(a)　$P(X = 5) = \binom{4}{2} \times (1 - 0.6)^2 \times 0.6^3 = 0.2074.$

　　　(b)　$P(X = 8) = \binom{7}{2} \times (1 - 0.6)^5 \times 0.6^3 = 0.0464.$

　　　(c)　$P(X \le 7) = P(X = 3) + P(X = 4) + P(X = 5) + P(X = 6) + P(X = 7)$
　　　　　$= 0.9037.$

　　　(d)　$P(X \ge 7) = 1 - P(X = 3) - P(X = 4) - P(X = 5) - P(X = 6) = 0.1792.$

3.2.4　Notice that a negative binomial distribution with parameters p and r can be thought of as the number of trials up to and including the rth success in a sequence of independent Bernoulli trials with a constant success probability p, which can be considered to be the number of trials up to and including the first success, plus the number of trials after the first success and up to and including the second success, plus the number of trials after the second success and up to and including the third success, and so on. Each of these r components has a geometric distribution with parameter p.

3.2.5　(a)　Consider a geometric distribution with parameter $p = 0.09$.
　　　　　$(1 - 0.09)^3 \times 0.09 = 0.0678.$

　　　(b)　Consider a negative binomial distribution with parameters $p = 0.09$ and $r = 3$.
　　　　　$\binom{9}{2} \times (1 - 0.09)^7 \times 0.09^3 = 0.0136.$

　　　(c)　$\frac{1}{0.09} = 11.11.$

　　　(d)　$\frac{3}{0.09} = 33.33.$

3.2.6　(a)　$\frac{1}{0.37} = 2.703.$

　　　(b)　$\frac{3}{0.37} = 8.108.$

　　　(c)　$P(X \le 10) = 0.7794$ where the random variable X has a negative binomial distribution with parameters $p = 0.37$ and $r = 3$.

　　　(d)　$P(X = 10) = \binom{9}{2} \times (1 - 0.37)^7 \times 0.37^3 = 0.0718.$

3.2.7   (a) Consider a geometric distribution with parameter $p = 0.25$.

$(1 - 0.25)^2 \times 0.25 = 0.1406$.

(b) Consider a negative binomial distribution with parameters $p = 0.25$ and $r = 4$.

$\binom{9}{3} \times (1 - 0.25)^6 \times 0.25^4 = 0.0584$.

The expected number of cards drawn before the fourth heart is obtained is the expectation of a negative binomial distribution with parameters $p = 0.25$ and $r = 4$ which is $\frac{4}{0.25} = 16$.

If the first two cards are spades then the probability that the first heart card is obtained on the fifth drawing is the same as the probability in part (a).

3.2.8   (a) $\frac{1}{0.77} = 1.299$

(b) Consider a geometric distribution with parameter $p = 0.23$.

$(1 - 0.23)^4 \times 0.23 = 0.0809$.

(c) Consider a negative binomial distribution with parameters $p = 0.77$ and $r = 3$.

$\binom{5}{2} \times (1 - 0.77)^3 \times 0.77^3 = 0.0555$.

(d) $P(B(8, 0.77) \geq 3) = 0.9973$.

3.3   The Hypergeometric Distribution

3.3.1   (a) $P(X = 4) = \dfrac{\dbinom{6}{4} \times \dbinom{5}{3}}{\dbinom{11}{7}} = \dfrac{5}{11}.$

(b) $P(X = 5) = \dfrac{\dbinom{6}{5} \times \dbinom{5}{2}}{\dbinom{11}{7}} = \dfrac{2}{11}.$

(c) $P(X \leq 3) = P(X = 2) + P(X = 3) = \dfrac{23}{66}.$

3.3.2

x_i	0	1	2	3	4	5
p_i	$\frac{3}{429}$	$\frac{40}{429}$	$\frac{140}{429}$	$\frac{168}{429}$	$\frac{70}{429}$	$\frac{8}{429}$

3.3.3   (a) $\dfrac{\dbinom{10}{3} \times \dbinom{7}{2}}{\dbinom{17}{5}} = \dfrac{90}{221}.$

(b) $\dfrac{\dbinom{10}{1} \times \dbinom{7}{4}}{\dbinom{17}{5}} = \dfrac{25}{442}.$

(c) $P(\text{no red balls}) + P(\text{one red ball}) + P(\text{two red balls}) = \dfrac{139}{442}.$

3.3.4 $\dfrac{\dbinom{16}{5} \times \dbinom{18}{7}}{\dbinom{34}{12}} = 0.2535.$

$P(B(12, 18/34) = 7) = 0.2131.$

3.3.5 $\dfrac{\dbinom{12}{3} \times \dbinom{40}{2}}{\dbinom{52}{5}} = \dfrac{55}{833}.$

The number of picture cards X in a hand of 13 cards has a hypergeometric distribution with $N = 52$, $n = 13$ and $r = 12$. The expected value is

$$E(X) \; = \; \frac{13 \times 12}{52} \; = \; 3$$

and the variance is

$$Var(X) \; = \; \left(\frac{52 - 13}{52 - 1}\right) \times 13 \times \frac{12}{52} \times \left(1 - \frac{12}{52}\right) \; = \; \frac{30}{17}.$$

3.3.6 $\quad \dfrac{\binom{4}{1} \times \binom{5}{2} \times \binom{6}{2}}{\binom{15}{5}} = \frac{200}{1001}.$

3.4 The Poisson Distribution

3.4.1   (a) $P(X = 1) = \dfrac{e^{-3.2} \times 3.2^1}{1!} = 0.1304.$

   (b) $P(X \leq 3) = P(X = 0) + P(X = 1) + P(X = 2) + P(X = 3) = 0.6025.$

   (c) $P(X \geq 6) = 1 - P(X \leq 5) = 0.1054.$

   (d) $P(X = 0 | X \leq 3) = \dfrac{P(X = 0)}{P(X \leq 3)} = \dfrac{0.0408}{0.6025} = 0.0677.$

3.4.2   (a) $P(X = 0) = \dfrac{e^{-2.1} \times 2.1^0}{0!} = 0.1225.$

   (b) $P(X \leq 2) = P(X = 0) + P(X = 1) + P(X = 2) = 0.6496.$

   (c) $P(X \geq 5) = 1 - P(X \leq 4) = 0.0621.$

   (d) $P(X = 1 | X \leq 2) = \dfrac{P(X = 1)}{P(X \leq 2)} = \dfrac{0.2572}{0.6496} = 0.3959.$

3.4.4 $P(X = 0) = \dfrac{e^{-2.4} \times 2.4^0}{0!} = 0.0907.$

$P(X \geq 4) = 1 - P(X \leq 3) = 0.2213.$

3.4.5   Use a Poisson distribution with $\lambda = \frac{25}{100} = 0.25.$

$$P(X = 0) = \dfrac{e^{-0.25} \times 0.25^0}{0!} = 0.7788.$$

$$P(X \leq 1) = P(X = 0) + P(X = 1) = 0.9735.$$

3.4.6   Use a Poisson distribution with $\lambda = 4.$

   (a) $P(X = 0) = \dfrac{e^{-4} \times 4^0}{0!} = 0.0183.$

   (b) $P(X \geq 6) = 1 - P(X \leq 5) = 0.2149.$

3.4.7 $P(B(500, 0.005) \leq 3)$ can be approximated as

$$P(Poisson(500 \times 0.005) \leq 3) \;=\; P(Poisson(2.5) \leq 3)$$

$$= \frac{e^{-2.5} \times 2.5^0}{0!} \;+\; \frac{e^{-2.5} \times 2.5^1}{1!} \;+\; \frac{e^{-2.5} \times 2.5^2}{2!} \;=\; 0.7576.$$

3.5 The Multinomial Distribution

3.5.1 (a) $$\frac{11!}{4! \times 5! \times 2!} \times 0.23^4 \times 0.48^5 \times 0.29^2 \;=\; 0.0416.$$

(b) $P(B(7, 0.23) < 3) = 0.7967.$

3.5.2 (a) $$\frac{15!}{3! \times 3! \times 9!} \times \left(\frac{1}{6}\right)^3 \times \left(\frac{1}{6}\right)^3 \times \left(\frac{2}{3}\right)^9 \;=\; 0.0558.$$

(b) $$\frac{15!}{3! \times 3! \times 4! \times 5!} \times \left(\frac{1}{6}\right)^3 \times \left(\frac{1}{6}\right)^3 \times \left(\frac{1}{6}\right)^4 \times \left(\frac{1}{2}\right)^5 \;=\; 0.0065.$$

(c) $$\frac{15!}{2! \times 13!} \times \left(\frac{1}{6}\right)^2 \times \left(\frac{5}{6}\right)^{13} \;=\; 0.2726.$$

The expected number of sixes is $\frac{15}{6} = 2.5$.

3.5.3 (a) $$\frac{8!}{2! \times 5! \times 1!} \times 0.09^2 \times 0.79^5 \times 0.12^1 \;=\; 0.0502.$$

(b) $$\frac{8!}{1! \times 5! \times 2!} \times 0.09^1 \times 0.79^5 \times 0.12^2 \;=\; 0.0670.$$

(c) $P(B(8, 0.09) \geq 2) = 0.1577.$

The expected number of misses is $8 \times 0.12 = 0.96$.

3.5.4 The expected number of dead seedlings is $22 \times 0.08 = 1.76$, the expected number of slow growth seedlings is $22 \times 0.19 = 4.18$, the expected number of medium growth seedlings is $22 \times 0.42 = 9.24$, and the expected number of strong growth seedlings is $22 \times 0.31 = 6.82$.

(a) $$\frac{22!}{3! \times 4! \times 6! \times 9!} \times 0.08^3 \times 0.19^4 \times 0.42^6 \times 0.31^9 \;=\; 0.0029.$$

(b) $$\frac{22!}{5! \times 5! \times 5! \times 7!} \times 0.08^5 \times 0.19^5 \times 0.42^5 \times 0.31^7 \;=\; 0.00038.$$

(c) $P(B(22, 0.08) \leq 2) = 0.7442.$

3.6   Supplementary Problems

3.6.1   (a) $P(B(18, 0.085) \geq 3) = 1 - P(B(18, 0.085) \leq 2) = 0.1931.$

(b) $P(B(18, 0.085) \leq 1) = 0.5401.$

(c) $18 \times 0.085 = 1.53.$

3.6.2 $P(B(13, 0.4) \geq 3) = 1 - P(B(13, 0.4) \leq 2) = 0.9421.$

The expected number of cells is $13 + (13 \times 0.4) = 18.2.$

3.6.3   (a) $\dfrac{8!}{2! \times 3! \times 3!} \times 0.40^2 \times 0.25^3 \times 0.35^3 = 0.0600.$

(b) $\dfrac{8!}{3! \times 1! \times 4!} \times 0.40^3 \times 0.25^1 \times 0.35^4 = 0.0672.$

(c) $P(B(8, 0.35) \leq 2) = 0.4278.$

3.6.4   (a) $P(X = 0) = \dfrac{e^{-2/3} \times (2/3)^0}{0!} = 0.5134.$

(b) $P(X = 1) = \dfrac{e^{-2/3} \times (2/3)^1}{1!} = 0.3423.$

(c) $P(X \geq 3) = 1 - P(X \leq 2) = 0.0302.$

3.6.5 $P(X = 2) = \dfrac{e^{-3.3} \times (3.3)^2}{2!} = 0.2008.$

$P(X \geq 6) = 1 - P(X \leq 5) = 0.1171.$

3.6.6   (a) Consider a negative binomial distribution with parameters $p = 0.55$ and $r = 4.$

(b) $P(X = 7) = \begin{pmatrix} 6 \\ 3 \end{pmatrix} \times (1 - 0.55)^3 \times 0.55^4 = 0.1668.$

(c) $P(X = 6) = \begin{pmatrix} 5 \\ 3 \end{pmatrix} \times (1 - 0.55)^2 \times 0.55^4 = 0.1853.$

(d) The probability that team A wins the series in game 5 is

$\begin{pmatrix} 4 \\ 3 \end{pmatrix} \times (1 - 0.55)^1 \times 0.55^4 = 0.1647.$

The probability that team B wins the series in game 5 is

$\begin{pmatrix} 4 \\ 3 \end{pmatrix} \times (1 - 0.45)^1 \times 0.45^4 = 0.0902.$

The probability that the series is over after game five is $0.1647 + 0.0902 = 0.2549.$

(e) The probability that team A wins the series in game 4 is $0.55^4 = 0.0915$.
The probability that team A wins the series is $0.0915 + 0.1647 + 0.1853 + 0.1668 = 0.6083$.

3.6.7 (a) Consider a negative binomial distribution with parameters $p = 0.58$ and $r = 3$.
$$P(X = 9) = \binom{8}{2} \times (1 - 0.58)^6 \times 0.58^3 = 0.0300.$$

(b) Consider a negative binomial distribution with parameters $p = 0.42$ and $r = 4$.
$$P(X \leq 7) = P(X = 4) + P(X = 5) + P(X = 6) + P(X = 7) = 0.3294.$$

3.6.8
$$P(two\ red\ balls|head) = \frac{\binom{6}{2} \times \binom{5}{1}}{\binom{11}{3}} = \frac{5}{11}.$$

$$P(two\ red\ balls|tail) = \frac{\binom{5}{2} \times \binom{6}{1}}{\binom{11}{3}} = \frac{4}{11}.$$

Then

$$P(two\ red\ balls) = (P(head) \times P(two\ red\ balls|head))$$

$$+ (P(tail) \times P(two\ red\ balls|tail))$$

$$= \left(0.5 \times \frac{5}{11}\right) + \left(0.5 \times \frac{4}{11}\right) = \frac{9}{22}$$

and

$$P(head|two\ red\ balls) = \frac{P(head\ and\ two\ red\ balls)}{P(two\ red\ balls)}$$

$$= \frac{P(head) \times P(two\ red\ balls|head)}{P(two\ red\ balls)} = \frac{5}{9}.$$

Chapter 4

Continuous Probability Distributions

4.1 The Uniform Distribution

4.1.1   (a) $E(X) = \frac{-3+8}{2} = 2.5$.

      (b) $\sigma = \frac{8-(-3)}{\sqrt{12}} = 3.175$.

      (c)   The upper quartile is 5.25.

      (d) $P(0 \leq X \leq 4) = \int_0^4 \frac{1}{11}\, dx = \frac{4}{11}$.

4.1.2   (a) $E(X) = \frac{1.43+1.60}{2} = 1.515$.

      (b) $\sigma = \frac{1.60-1.43}{\sqrt{12}} = 0.0491$.

      (c) $F(x) = \frac{x-1.43}{1.60-1.43} = \frac{x-1.43}{0.17}$ for $1.43 \leq x \leq 1.60$.

      (d) $F(1.48) = \frac{1.48-1.43}{0.17} = \frac{0.05}{0.17} = 0.2941$.

      (e) $F(1.5) = \frac{1.5-1.43}{0.17} = \frac{0.07}{0.17} = 0.412$.

      The number of batteries with a voltage less than 1.5 Volts has a binomial distribution with parameters $n = 50$ and $p = 0.412$ so that the expected value is

$$E(X) = n \times p = 50 \times 0.412 = 20.6$$

and the variance is

$$Var(X) = n \times p \times (1-p) = 50 \times 0.412 \times 0.588 = 12.11.$$

4.1.3   (a) These four intervals have probabilities 0.30, 0.20, 0.25, and 0.25 respectively and
              the expectations and variances are calculated from the binomial distribution.

              The expectations are
              $20 \times 0.30 = 6$
              $20 \times 0.20 = 4$
              $20 \times 0.25 = 5$ and
              $20 \times 0.25 = 5$.

              The variances are
              $20 \times 0.30 \times 0.70 = 4.2$
              $20 \times 0.20 \times 0.80 = 3.2$
              $20 \times 0.25 \times 0.75 = 3.75$ and
              $20 \times 0.25 \times 0.75 = 3.75$.

        (b) Using the multinomial distribution the probability is

              $$\frac{20!}{5! \times 5! \times 5! \times 5!} \; 0.30^5 \times 0.20^5 \times 0.25^5 \times 0.25^5 \; = \; 0.0087.$$

4.1.4   (a) $E(X) = \frac{0.0 + 2.5}{2} = 1.25$.

              $Var(X) = \frac{(2.5 - 0.0)^2}{12} = 0.5208$.

        (b) The probability that a piece of scrapwood is longer than 1 meter is $\frac{1.5}{2.5} = 0.6$.
              $P(B(25, 0.6) \geq 20) = 0.0294$.

4.2 The Exponential Distribution

4.2.2   (a) $E(X) = \frac{1}{0.1} = 10$.

(b) $P(X \geq 10) = 1 - F(10) = 1 - (1 - e^{-0.1 \times 10}) = e^{-1} = 0.3679$.

(c) $P(X \leq 5) = F(5) = 1 - e^{-0.1 \times 5} = 0.3935$.

(d) The *additional* waiting time also has an exponential distribution with parameter $\lambda = 0.1$. The probability that the total waiting time is longer than 15 minutes is the probability that the *additional* waiting time is longer than 10 minutes which is 0.3679 from part (b).

(e) $E(X) = \frac{0 + 20}{2} = 10$ as in the previous case. However, in this case the *additional* waiting time has a $U(0, 15)$ distribution.

4.2.3   (a) $E(X) = \frac{1}{0.2} = 5$.

(b) $\sigma = \frac{1}{0.2} = 5$.

(c) The median is $\frac{0.693}{0.2} = 3.47$.

(d) $P(X \geq 7) = 1 - F(7) = 1 - (1 - e^{-0.2 \times 7}) = e^{-1.4} = 0.2466$.

(e) The memoryless property of the exponential distribution implies that the required probability is
$P(X \geq 2) = 1 - F(2) = 1 - (1 - e^{-0.2 \times 2}) = e^{-0.4} = 0.6703$.

4.2.4   (a) $P(X \leq 5) = F(5) = 1 - e^{-0.31 \times 5} = 0.7878$.

(b) Consider a binomial distribution with parameters $n = 12$ and $p = 0.7878$. The expected value is
$$E(X) \;=\; n \times p \;=\; 12 \times 0.7878 \;=\; 9.45$$
and the variance is
$$Var(X) \;=\; n \times p \times (1 - p) \;=\; 12 \times 0.7878 \times 0.2122 \;=\; 2.01.$$

(c) $P(B(12, 0.7878) \leq 9) = 0.4845$.

4.2.5
$$F(x) \;=\; \int_{-\infty}^{x} \frac{1}{2} \lambda e^{-\lambda(\theta - y)} \, dy \;=\; \frac{1}{2} e^{-\lambda(\theta - x)}$$

for $-\infty \leq x \leq \theta$, and

$$F(x) \;=\; \frac{1}{2} + \int_{\theta}^{x} \frac{1}{2} \lambda e^{-\lambda(y - \theta)} \, dy \;=\; 1 - \frac{1}{2} e^{-\lambda(x - \theta)}$$

for $\theta \leq x \leq \infty$.

(a) $P(X \leq 0) = F(0) = \frac{1}{2}e^{-3(2-0)} = 0.0012$.

(b) $P(X \geq 1) = 1 - F(1) = 1 - \frac{1}{2}e^{-3(2-1)} = 0.9751$.

4.2.6   (a) $E(X) = \frac{1}{2} = 0.5$.

(b) $P(X \geq 1) = 1 - F(1) = 1 - (1 - e^{-2\times 1}) = e^{-2} = 0.1353$.

(c) A Poisson distribution with parameter $2 \times 3 = 6$.

(d) $P(X \leq 4) = P(X = 0) + P(X = 1) + P(X = 2) + P(X = 3) + P(X = 4)$

$$= \frac{e^{-6} \times 6^0}{0!} + \frac{e^{-6} \times 6^1}{1!} + \frac{e^{-6} \times 6^2}{2!} + \frac{e^{-6} \times 6^3}{3!} + \frac{e^{-6} \times 6^4}{4!} = 0.2851.$$

4.2.7   (a) $\lambda = 1.8$.

(b) $E(X) = \frac{1}{1.8} = 0.5556$.

(c) $P(X \geq 1) = 1 - F(1) = 1 - (1 - e^{-1.8\times 1}) = e^{-1.8} = 0.1653$.

(d) A Poisson distribution with parameter $1.8 \times 4 = 7.2$.

(e) $P(X \geq 4) = 1 - P(X = 0) - P(X = 1) - P(X = 2) - P(X = 3)$

$$= 1 - \frac{e^{-7.2} \times 7.2^0}{0!} - \frac{e^{-7.2} \times 7.2^1}{1!} - \frac{e^{-7.2} \times 7.2^2}{2!} - \frac{e^{-7.2} \times 7.2^3}{3!} = 0.9281.$$

4.3 The Gamma Distribution

4.3.1 $\Gamma(5.5) = 4.5 \times 3.5 \times 2.5 \times 1.5 \times 0.5 \times \sqrt{\pi} = 52.34$.

4.3.3   (a) $f(3) = 0.2055$, $F(3) = 0.3823$, and $F^{-1}(0.5) = 3.5919$.

   (b) $f(3) = 0.0227$, $F(3) = 0.9931$, and $F^{-1}(0.5) = 1.3527$.

   (c) $f(3) = 0.2592$, $F(3) = 0.6046$, and $F^{-1}(0.5) = 2.6229$.
   In this case

$$f(3) = \frac{1.4^4 \times 3^{4-1} \times e^{-1.4 \times 3}}{3!} = 0.2592.$$

4.3.4   (a) $E(X) = \frac{5}{0.9} = 5.556$.

   (b) $\sigma = \frac{\sqrt{5}}{0.9} = 2.485$.

   (c) From the computer the lower quartile is $F^{-1}(0.25) = 3.743$ and the upper quartile is $F^{-1}(0.75) = 6.972$.

   (d) From the computer $P(X \geq 6) = 0.3733$.

4.3.5   (a) A gamma distribution with parameters $k = 4$ and $\lambda = 2$.

   (b) $E(X) = \frac{4}{2} = 2$.

   (c) $\sigma = \frac{\sqrt{4}}{2} = 1$.

   (d) $P(X \geq 3) = 0.1512$ where the random variable X has a gamma distribution with parameters $k = 4$ and $\lambda = 2$.
   Also, $P(Y \leq 3) = 0.1512$ where the random variable Y has a Poisson distribution with parameter $2 \times 3 = 6$ and measures the number of imperfections in a 3 meter length of fiber.

4.3.6   (a) A gamma distribution with parameters $k = 3$ and $\lambda = 1.8$.

   (b) $E(X) = \frac{3}{1.8} = 1.667$.

   (c) $Var(X) = \frac{3}{1.8^2} = 0.9259$.

(d) $P(X \geq 3) = 0.0948$ where the random variable X has a gamma distribution with parameters $k = 3$ and $\lambda = 1.8$.

Also, $P(Y \leq 2) = 0.0948$ where the random variable Y has a Poisson distribution with parameter $1.8 \times 3 = 5.4$ and measures the number of arrivals in a 3 hour period.

4.4 The Weibull Distribution

4.4.2 (a) $\dfrac{(-\ln(1-0.5))^{1/4.9}}{0.22} = 4.218.$

(b) $\dfrac{(-\ln(1-0.75))^{1/4.9}}{0.22} = 4.859.$

$\dfrac{(-\ln(1-0.25))^{1/4.9}}{0.22} = 3.525.$

(c) $F(x) = 1 - e^{-(0.22x)^{4.9}}.$

$P(2 \le X \le 7) = F(7) - F(2) = 0.9820.$

4.4.3 (a) $\dfrac{(-\ln(1-0.5))^{1/2.3}}{1.7} = 0.5016.$

(b) $\dfrac{(-\ln(1-0.75))^{1/2.3}}{1.7} = 0.6780.$

$\dfrac{(-\ln(1-0.25))^{1/2.3}}{1.7} = 0.3422.$

(c) $F(x) = 1 - e^{-(1.7x)^{2.3}}.$

$P(0.5 \le X \le 1.5) = F(1.5) - F(0.5) = 0.5023.$

4.4.4 (a) $\dfrac{(-\ln(1-0.5))^{1/0.5}}{0.5} = 0.9609.$

(b) $\dfrac{(-\ln(1-0.01))^{1/0.5}}{0.5} = 0.0002.$

(c) $E(X) = \dfrac{1}{0.5}\,\Gamma\left(1 + \dfrac{1}{0.5}\right) = 4.$

$Var(X) = \dfrac{1}{0.5^2}\left\{\Gamma\left(1 + \dfrac{2}{0.5}\right) - \Gamma\left(1 + \dfrac{1}{0.5}\right)^2\right\} = 80.$

(d) $P(X \le 3) = F(3) = 1 - e^{-(0.5\times3)^{0.5}} = 0.7062.$

The probability that at least one circuit is working after three hours is $1 - 0.7062^3 = 0.6479.$

4.4.5   (a) $\dfrac{(-\ln(1-0.5))^{1/0.4}}{0.5} = 0.8000.$

(b) $\dfrac{(-\ln(1-0.75))^{1/0.4}}{0.5} = 4.5255.$

$\dfrac{(-\ln(1-0.25))^{1/0.4}}{0.5} = 0.0888.$

(c) $\dfrac{(-\ln(1-0.95))^{1/0.4}}{0.5} = 31.066.$

$\dfrac{(-\ln(1-0.99))^{1/0.4}}{0.5} = 91.022.$

(d) $F(x) = 1 - e^{-(0.5x)^{0.4}}.$

$P(3 \leq X \leq 5) = F(5) - F(3) = 0.0722.$

4.4.6   (a) $\dfrac{(-\ln(1-0.5))^{1/1.5}}{0.03} = 26.11.$

$\dfrac{(-\ln(1-0.75))^{1/1.5}}{0.03} = 41.44.$

$\dfrac{(-\ln(1-0.99))^{1/1.5}}{0.03} = 92.27.$

(b) $F(x) = 1 - e^{-(0.03x)^{1.5}}.$

$P(X \geq 30) = 1 - F(30) = 0.4258.$

The number of ants still alive after 30 minutes has a binomial distribution with parameters $n = 500$ and $p = 0.4258$. The expected value is

$E(X) = n \times p = 500 \times 0.4258 = 212.9$

and the variance is

$Var(X) = n \times p \times (1-p) = 500 \times 0.4258 \times 0.5742 = 122.2.$

4.5 The Beta Distribution

4.5.1 (a) $\quad \displaystyle\int_0^1 A\,x^3(1-x)^2\,dx \;=\; 1 \quad\Rightarrow\quad A \;=\; 60.$

(b) $\quad E(X) \;=\; \displaystyle\int_0^1 60\,x^4(1-x)^2\,dx \;=\; \frac{4}{7}.$

$\quad E(X^2) \;=\; \displaystyle\int_0^1 60\,x^5(1-x)^2\,dx \;=\; \frac{5}{14}.$

Therefore $Var(X) = \frac{5}{14} - \left(\frac{4}{7}\right)^2 = \frac{3}{98}.$

(c) This is a beta distribution with $a = 4$ and $b = 3$.

$$E(X) \;=\; \frac{4}{4+3} \;=\; \frac{4}{7}.$$

$$Var(X) \;=\; \frac{4 \times 3}{(4+3)^2 \times (4+3+1)} \;=\; \frac{3}{98}.$$

4.5.2 (a) This is a beta distribution with $a = 10$ and $b = 4$.

(b) $\quad A \;=\; \dfrac{\Gamma(10+4)}{\Gamma(10)\times\Gamma(4)} \;=\; \dfrac{13!}{9! \times 3!} \;=\; 2,860.$

(c) $\quad E(X) \;=\; \dfrac{10}{10+4} \;=\; \dfrac{5}{7}.$

(d) $\quad Var(X) \;=\; \dfrac{10 \times 4}{(10+4)^2 \times (10+4+1)} \;=\; \dfrac{2}{147}$

and $\sigma = \sqrt{\frac{2}{147}} = 0.1166.$

(e) $\quad F(x) \;=\; \displaystyle\int_0^x 2,860\,y^9\,(1-y)^3\,dy$

$$= \; 2,860\left(\frac{x^{10}}{10} - \frac{3x^{11}}{11} + \frac{x^{12}}{4} - \frac{x^{13}}{13}\right) \quad\text{for}\quad 0 \le x \le 1.$$

4.5.3 (a) $f(0.5) = 1.9418$, $F(0.5) = 0.6753$, and $F^{-1}(0.5) = 0.5406.$
(b) $f(0.5) = 0.7398$, $F(0.5) = 0.7823$, and $F^{-1}(0.5) = 0.4579.$
(c) $f(0.5) = 0.6563$, $F(0.5) = 0.9375$, and $F^{-1}(0.5) = 0.3407.$
In this case

$$f(0.5) \;=\; \frac{\Gamma(2+6)}{\Gamma(2)\times\Gamma(6)} \times 0.5^{2-1} \times (1-0.5)^{6-1} \;=\; 0.65625.$$

4.5.4   (a) $3 \leq y \leq 7$.

    (b) $E(X) = \dfrac{2.1}{2.1 + 2.1} = \dfrac{1}{2}.$

Therefore, $E(Y) = 3 + (4 \times E(X)) = 5$.

$$Var(X) = \frac{2.1 \times 2.1}{(2.1 + 2.1)^2 \times (2.1 + 2.1 + 1)} = 0.0481.$$

Therefore, $Var(Y) = 4^2 \times Var(X) = 0.1923$.

    (c) $P(Y \leq 5) = P(X \leq 0.5) = 0.5$ since the random variable X has a symmetric beta distribution.

4.5.5   (a) $E(X) = \dfrac{7.2}{7.2 + 2.3} = 0.7579.$

$$Var(X) = \frac{7.2 \times 2.3}{(7.2 + 2.3)^2 \times (7.2 + 2.3 + 1)} = 0.0175.$$

    (b) From the computer $P(X \geq 0.9) = 0.1368$.

4.5.6   (a) $E(X) = \dfrac{8.2}{8.2 + 11.7} = 0.4121.$

    (b) $Var(X) = \dfrac{8.2 \times 11.7}{(8.2 + 11.7)^2 \times (8.2 + 11.7 + 1)} = 0.0116$

and $\sigma = \sqrt{0.0116} = 0.1077$.

    (c) From the computer $F^{-1}(0.5) = 0.4091$.

4.6   Supplementary Problems

4.6.1 $F(0) = P(winnings = 0) = \frac{1}{4}$,

$F(x) = P(winnings \leq x) = \frac{1}{4} + \frac{x}{720}$ for $0 \leq x \leq 360$,

$F(x) = P(winnings \leq x) = \frac{\sqrt{x+72,540}}{360}$ for $360 \leq x \leq 57,060$,

$F(x) = 1$ for $57,060 \leq x$.

4.6.2   (a) $\frac{0.693}{\lambda} = 1.5$ $\Rightarrow$ $\lambda = 0.462$.

(b) $P(X \geq 2) = 1 - F(2) = 1 - (1 - e^{-0.462 \times 2}) = e^{-0.924} = 0.397$.

$P(X \leq 1) = F(1) = 1 - e^{-0.462 \times 1} = 0.370$.

4.6.3   (a) $E(X) = \frac{1}{0.7} = 1.4286$.

(b) $P(X \geq 3) = 1 - F(3) = 1 - (1 - e^{-0.7 \times 3}) = e^{-2.1} = 0.1225$.

(c) $\frac{0.693}{0.7} = 0.9902$.

(d)   A Poisson distribution with parameter $0.7 \times 10 = 7$.

(e) $P(X \geq 5) = 1 - P(X = 0) - P(X = 1) - P(X = 2) - P(X = 3) - P(X = 4)$
$= 0.8270$.

(f)   A gamma distribution with parameters $k = 10$ and $\lambda = 0.7$.

$E(X) = \frac{10}{0.7} = 14.286$.

4.6.4   (a) $E(X) = \frac{1}{5.2} = 0.1923$.

(b) $P(X \leq 1/6) = F(1/6) = 1 - e^{-5.2 \times 1/6} = 0.5796$.

(c)   A gamma distribution with parameters $k = 10$ and $\lambda = 5.2$.

(d) $E(X) = \frac{10}{5.2} = 1.923$.

(e) $P(X > 5) = 0.4191$ where the random variable X has a Poisson distribution
with parameter 5.2.

4.6.5   (a)   The total area under the triangle is one so the height at the midpoint is $\frac{2}{b-a}$.

(b) $P(X \leq a/4 + 3b/4) = P(X \leq a + 3(b-a)/4) = \frac{7}{8}$.

(c) $Var(X) = \frac{(b-a)^2}{24}$.

(d) $F(x) = \frac{2(x-a)^2}{(b-a)^2}$ for $a \leq x \leq \frac{a+b}{2}$,

$F(x) = 1 - \frac{2(b-x)^2}{(b-a)^2}$ for $\frac{a+b}{2} \leq x \leq b$.

4.6.6   (a) $\frac{(-\ln(1-0.5))^{1/0.2}}{8.2} = 0.0195$.

$\frac{(-\ln(1-0.75))^{1/0.2}}{8.2} = 0.6244$.

$\frac{(-\ln(1-0.95))^{1/0.2}}{8.2} = 29.42$.

(b) $E(X) = \frac{1}{8.2} \Gamma\left(1 + \frac{1}{0.2}\right) = 14.63$.

$Var(X) = \frac{1}{8.2^2} \left\{ \Gamma\left(1 + \frac{2}{0.2}\right) - \Gamma\left(1 + \frac{1}{0.2}\right)^2 \right\} = 53,753$.

(c) $F(x) = 1 - e^{-(8.2x)^{0.2}}$.

$P(10 \leq X \leq 20) = F(20) - F(10) = 0.0270$.

4.6.7   (a) $E(X) = \frac{2.7}{2.7 + 2.9} = 0.4821$.

(b) $Var(X) = \frac{2.7 \times 2.9}{(2.7 + 2.9)^2 \times (2.7 + 2.9 + 1)} = 0.0378$

and $\sigma = \sqrt{0.0378} = 0.1945$.

(c) From the computer $P(X \geq 0.5) = 0.4637$.

Chapter 5

The Normal Distribution

5.1 Probability Calculations using the Normal Distribution

5.1.1   (a) $\Phi(1.34) = 0.9099$.

       (b) $1 - \Phi(-0.22) = 0.5871$.

       (c) $\Phi(0.43) - \Phi(-2.19) = 0.6521$.

       (d) $\Phi(1.76) - \Phi(0.09) = 0.4249$.

       (e) $\Phi(0.38) - \Phi(-0.38) = 0.2960$.

       (f) Solving $\Phi(x) = 0.55$ gives $x = 0.1257$.

       (g) Solving $1 - \Phi(x) = 0.72$ gives $x = -0.5828$.

       (h) Solving $\Phi(x) - \Phi(-x) = (2 \times \Phi(x)) - 1 = 0.31$ gives $x = 0.3989$.

5.1.2   (a) $\Phi(-0.77) = 0.2206$.

       (b) $1 - \Phi(0.32) = 0.3745$.

       (c) $\Phi(-1.59) - \Phi(-3.09) = 0.0549$.

       (d) $\Phi(1.80) - \Phi(-0.82) = 0.7580$.

       (e) $1 - (\Phi(0.91) - \Phi(-0.91)) = 0.3628$.

       (f) Solving $\Phi(x) = 0.23$ gives $x = -0.7388$.

       (g) Solving $1 - \Phi(x) = 0.51$ gives $x = -0.0251$.

(h) Solving $1 - (\Phi(x) - \Phi(-x)) = 2 - (2 \times \Phi(x)) = 0.42$ gives $x = 0.8064$.

5.1.3   (a) $P(X \leq 10.34) = \Phi(\frac{10.34-10}{\sqrt{2}}) = 0.5950$.

(b) $P(X \geq 11.98) = 1 - \Phi(\frac{11.98-10}{\sqrt{2}}) = 0.0807$.

(c) $P(7.67 \leq X \leq 9.90) = \Phi(\frac{9.90-10}{\sqrt{2}}) - \Phi(\frac{7.67-10}{\sqrt{2}}) = 0.4221$.

(d) $P(10.88 \leq X \leq 13.22) = \Phi(\frac{13.22-10}{\sqrt{2}}) - \Phi(\frac{10.88-10}{\sqrt{2}}) = 0.2555$.

(e) $P(|X - 10| \leq 3) = P(7 \leq X \leq 13) = \Phi(\frac{13-10}{\sqrt{2}}) - \Phi(\frac{7-10}{\sqrt{2}}) = 0.9662$.

(f) Solving $P(N(10, 2^2) \leq x) = 0.81$ gives $x = 11.2415$.

(g) Solving $P(N(10, 2^2) \geq x) = 0.04$ gives $x = 12.4758$.

(h) Solving $P(|N(10, 2^2) - 10| \geq x) = 0.63$ gives $x = 0.6812$.

5.1.4   (a) $P(X \leq 0) = \Phi(\frac{0-(-7)}{\sqrt{14}}) = 0.9693$.

(b) $P(X \geq -10) = 1 - \Phi(\frac{-10-(-7)}{\sqrt{14}}) = 0.7887$.

(c) $P(-15 \leq X \leq -1) = \Phi(\frac{-1-(-7)}{\sqrt{14}}) - \Phi(\frac{-15-(-7)}{\sqrt{14}}) = 0.9293$.

(d) $P(-5 \leq X \leq 2) = \Phi(\frac{2-(-7)}{\sqrt{14}}) - \Phi(\frac{-5-(-7)}{\sqrt{14}}) = 0.2884$.

(e) $P(|X + 7| \geq 8) = 1 - P(-15 \leq X \leq 1) = 1 - (\Phi(\frac{1-(-7)}{\sqrt{14}}) - \Phi(\frac{-15-(-7)}{\sqrt{14}}))$
$= 0.0326$.

(f) Solving $P(N(-7, 14^2) \leq x) = 0.75$ gives $x = -4.4763$.

(g) Solving $P(N(-7, 14^2) \geq x) = 0.27$ gives $x = -4.7071$.

(h) Solving $P(|N(-7, 14^2) + 7| \leq x) = 0.44$ gives $x = 3.8192$.

5.1.5   Solving $P(X \leq 5) = \Phi(\frac{5-\mu}{\sigma}) = 0.8$ and $P(X \geq 0) = 1 - \Phi(\frac{0-\mu}{\sigma}) = 0.6$ gives
$\mu = 1.1569$ and $\sigma = 4.5663$.

5.1.6 Solving $P(X \leq 10) = \Phi(\frac{10-\mu}{\sigma}) = 0.55$ and $P(X \leq 0) = \Phi(\frac{0-\mu}{\sigma}) = 0.4$ gives
$\mu = 6.6845$ and $\sigma = 26.3845$.

5.1.7 $P(X \leq \mu + \sigma z_\alpha) = \Phi(\frac{\mu + \sigma z_\alpha - \mu}{\sigma}) = \Phi(z_\alpha) = 1 - \alpha.$

$P(\mu - \sigma z_{\alpha/2} \leq X \leq \mu + \sigma z_{\alpha/2}) = \Phi(\frac{\mu + \sigma z_{\alpha/2} - \mu}{\sigma}) - \Phi(\frac{\mu - \sigma z_{\alpha/2} - \mu}{\sigma}) = \Phi(z_{\alpha/2}) - \Phi(-z_{\alpha/2})$
$= 1 - \alpha/2 - \alpha/2 = 1 - \alpha.$

5.1.8 Solving $\Phi(x) = 0.75$ gives $x = 0.6745$.

Solving $\Phi(x) = 0.25$ gives $x = -0.6745$.

The interquartile range of a $N(0,1)$ distribution is $0.6745 - (-0.6745) = 1.3490$.

The interquartile range of a $N(\mu, \sigma^2)$ distribution is $1.3490 \times \sigma$.

5.1.9 (a) $P(N(3.00, 0.12^2) \geq 3.2) = 0.0478$.

(b) $P(N(3.00, 0.12^2) \leq 2.7) = 0.0062$.

(c) Solving $P(3.00 - c \leq N(3.00, 0.12^2) \leq 3.00 + c) = 0.99$ gives $c = 0.12 \times z_{0.005}$
$= 0.12 \times 2.5758 = 0.3091$.

5.1.10 (a) $P(N(1.03, 0.014^2) \leq 1) = 0.0161$.

(b) $P(N(1.05, 0.016^2) \leq 1) = 0.0009$.
There is a decrease in the proportion of underweight packets.

(c) The expected excess weight is $\mu - 1$ which is 0.03 and 0.05.

5.1.11 (a) Solving $P(N(4.3, 0.12^2) \leq x) = 0.75$ gives $x = 4.3809$.
Solving $P(N(4.3, 0.12^2) \leq x) = 0.25$ gives $x = 4.2191$.

(b) Solving $P(4.3 - c \leq N(4.3, 0.12^2) \leq 4.3 + c) = 0.80$ gives
$c = 0.12 \times z_{0.10} = 0.12 \times 1.2816 = 0.1538$.

5.1.12 (a) $P(N(0.0046, 9.6 \times 10^{-8}) \leq 0.005) = 0.9017$.

(b) $P(0.004 \leq N(0.0046, 9.6 \times 10^{-8}) \leq 0.005) = 0.8753$.

(c) Solving $P(N(0.0046, 9.6 \times 10^{-8}) \leq x) = 0.10$ gives $x = 0.0042$.

(d) Solving $P(N(0.0046, 9.6 \times 10^{-8}) \leq x) = 0.99$ gives $x = 0.0053$.

5.1.13   (a) $P(N(23.8, 1.28) \leq 23.0) = 0.2398$.

(b) $P(N(23.8, 1.28) \geq 24.0) = 0.4298$.

(c) $P(24.2 \leq N(23.8, 1.28) \leq 24.5) = 0.0937$.

(d) Solving $P(N(23.8, 1.28) \leq x) = 0.75$ gives $x = 24.56$.

(e) Solving $P(N(23.8, 1.28) \leq x) = 0.95$ gives $x = 25.66$.

5.1.14   Solving $P(N(\mu, 0.05^2) \leq 10) = 0.01$ gives
$\mu = 10 + (0.05 \times z_{0.01}) = 10 + (0.05 \times 2.3263) = 10.1163$.

5.2 Linear Combinations of Normal Random Variables

5.2.1   (a) $P(N(3.2 + (-2.1), 6.5 + 3.5) \geq 0) = 0.6360$.

         (b) $P(N(3.2 + (-2.1) - (2 \times 12.0), 6.5 + 3.5 + (2^2 \times 7.5)) \leq -20) = 0.6767$.

         (c) $P(N((3 \times 3.2) + (5 \times (-2.1)), (3^2 \times 6.5) + (5^2 \times 3.5)) \geq 1) = 0.4375$.

         (d) $P(N((4 \times 3.2) - (4 \times (-2.1)) + (2 \times 12.0), (4^2 \times 6.5) + (4^2 \times 3.5) + (2^2 \times 7.5))$
            $\leq 25) = 0.0714$.

         (e) $P(|N(3.2 + (6 \times (-2.1)) + 12.0, 6.5 + (6^2 \times 3.5) + 7.5)| \geq 2) = 0.8689$.

         (f) $P(|N((2 \times 3.2) - (-2.1) - 6, (2^2 \times 6.5) + 3.5)| \leq 1) = 0.1315$.

5.2.2   (a) $P(N(-1.9 - 3.3, 2.2 + 1.7) \geq -3) = 0.1326$.

         (b) $P(N((2 \times (-1.9)) + (3 \times 3.3) + (4 \times 0.8), (2^2 \times 2.2) + (3^2 \times 1.7) + (4^2 \times 0.2))$
            $\leq 10) = 0.5533$.

         (c) $P(N((3 \times 3.3) - 0.8, (3^2 \times 1.7) + 0.2) \leq 8) = 0.3900$.

         (d) $P(N((2 \times (-1.9)) - (2 \times 3.3) + (3 \times 0.8), (2^2 \times 2.2) + (2^2 \times 1.7) + (3^2 \times 0.2))$
            $\leq -6) = 0.6842$.

         (e) $P(|N(-1.9 + 3.3 - 0.8, 2.2 + 1.7 + 0.2)| \geq 1.5) = 0.4781$.

         (f) $P(|N((4 \times (-1.9)) - 3.3 + 10, (4^2 \times 2.2) + 1.7)| \leq 0.5) = 0.0648$.

5.2.3   (a) $\Phi(0.5) - \Phi(-0.5) = 0.3830$.

         (b) $P(|N(0, \frac{1}{8})| \leq 0.5) = 0.8428$.

         (c)   Need $0.5\sqrt{n} \geq z_{0.005} = 2.5758$ which is satisfied for $n \geq 27$.

5.2.4   (a) $N(4.3 + 4.3, 0.12^2 + 0.12^2) = N(8.6, 0.0288)$.

         (b) $N(4.3, \frac{0.12^2}{12}) = N(4.3, 0.0012)$.

(c) Need $z_{0.0015} \times \frac{0.12}{\sqrt{n}} = 2.9677 \times \frac{0.12}{\sqrt{n}} \leq 0.05$ which is satisfied for $n \geq 51$.

5.2.5 $P(144 \leq N(37 + 37 + 24 + 24 + 24, 0.49 + 0.49 + 0.09 + 0.09 + 0.09) \leq 147) = 0.7777$.

5.2.6   (a) $Var(Y) = (p^2 \times \sigma_1^2) + ((1 - p)^2 \times \sigma_2^2)$.
        The minimum variance is
        $$\frac{1}{\frac{1}{\sigma_1^2} + \frac{1}{\sigma_2^2}} = \frac{\sigma_1^2 \, \sigma_2^2}{\sigma_1^2 + \sigma_2^2}.$$

        (b) In this case
        $$Var(Y) = \sum_{i=1}^{n} p_i^2 \, \sigma_i^2.$$

        The variance is minimized with
        $$p_i = \frac{\frac{1}{\sigma_i^2}}{\frac{1}{\sigma_1^2} + \ldots + \frac{1}{\sigma_n^2}}$$

        and the minimum variance is
        $$\frac{1}{\frac{1}{\sigma_1^2} + \ldots + \frac{1}{\sigma_n^2}}.$$

5.2.7   (a) $1.05y + 1.05(1000 - y) = \$1050$.

        (b) $0.0002y^2 + 0.0003(1000 - y)^2$.

        (c) The variance is minimized with $y = 600$ and the minimum variance is 120.
             $P(N(1050, 120) \geq 1060) = 0.1807$.

5.2.8   (a) $P(N(3.00 + 3.00 + 3.00, 0.12^2 + 0.12^2 + 0.12^2) \geq 9.50) = 0.0081$.

        (b) $P(N(3.00, \frac{0.12^2}{7}) \leq 3.10) = 0.9863$.

5.2.9   (a) $N(22 \times 1.03, 22 \times 0.014^2) = N(22.66, 4.312 \times 10^{-3})$.

        (b) Solving $P(N(22.66, 4.312 \times 10^{-3}) \leq x) = 0.75$ gives $x = 22.704$.
             Solving $P(N(22.66, 4.312 \times 10^{-3}) \leq x) = 0.25$ gives $x = 22.616$.

5.3 Approximating Distributions with the Normal Distribution

5.3.1   (a)   The exact probability is 0.3823.

         The normal approximation is $1 - \Phi(\frac{8-0.5-(10\times0.7)}{\sqrt{10\times0.7\times0.3}}) = 0.3650.$

      (b)   The exact probability is 0.9147.

         The normal approximation is $\Phi(\frac{7+0.5-(15\times0.3)}{\sqrt{15\times0.3\times0.7}}) - \Phi(\frac{1+0.5-(15\times0.3)}{\sqrt{15\times0.3\times0.7}}) = 0.9090.$

      (c)   The exact probability is 0.7334.

         The normal approximation is $\Phi(\frac{4+0.5-(9\times0.4)}{\sqrt{9\times0.4\times0.6}}) = 0.7299.$

      (d)   The exact probability is 0.6527.

         The normal approximation is $\Phi(\frac{11+0.5-(14\times0.6)}{\sqrt{14\times0.6\times0.4}}) - \Phi(\frac{7+0.5-(14\times0.6)}{\sqrt{14\times0.6\times0.4}}) = 0.6429.$

5.3.2   (a)   The exact probability is 0.0106.

         The normal approximation is $1 - \Phi(\frac{7-0.5-(10\times0.3)}{\sqrt{10\times0.3\times0.7}}) = 0.0079.$

      (b)   The exact probability is 0.6160.

         The normal approximation is $\Phi(\frac{12+0.5-(21\times0.5)}{\sqrt{21\times0.5\times0.5}}) - \Phi(\frac{8+0.5-(21\times0.5)}{\sqrt{21\times0.5\times0.5}}) = 0.6172.$

      (c)   The exact probability is 0.9667.

         The normal approximation is $\Phi(\frac{3+0.5-(7\times0.2)}{\sqrt{7\times0.2\times0.8}}) = 0.9764.$

      (d)   The exact probability is 0.3410.

         The normal approximation is $\Phi(\frac{11+0.5-(12\times0.65)}{\sqrt{12\times0.65\times0.35}}) - \Phi(\frac{8+0.5-(12\times0.65)}{\sqrt{12\times0.65\times0.35}}) = 0.3233.$

5.3.3   The required probability is

$$\Phi\left(0.02\sqrt{n} + \frac{1}{\sqrt{n}}\right) - \Phi\left(-0.02\sqrt{n} - \frac{1}{\sqrt{n}}\right)$$

which is equal to
0.2358 for $n = 100$,
0.2764 for $n = 200$,
0.3772 for $n = 500$,
0.4934 for $n = 1,000$, and
0.6408 for $n = 2,000$.

5.3.4   (a) $\Phi(\frac{180+0.5-(1,000\times1/6)}{\sqrt{1,000\times1/6\times5/6}}) - \Phi(\frac{149+0.5-(1,000\times1/6)}{\sqrt{1,000\times1/6\times5/6}}) = 0.8072.$

      (b) It is required that

$$1 - \Phi\left(\frac{50 - 0.5 - n/6}{\sqrt{n \times 1/6 \times 5/6}}\right) \geq 0.99$$

      which is satisfied for $n \geq 402$.

5.3.5   (a) A normal distribution can be used with $\mu = 500 \times 2.4 = 1,200$ and $\sigma^2 = 500 \times 2.4 = 1,200$.

      (b) $P(N(1200, 1200) \geq 1250) = 0.0745.$

5.3.6   The normal approximation is $1 - \Phi(\frac{150-0.5-(15,000\times1/125)}{\sqrt{15,000\times1/125\times124/125}}) = 0.0034.$

5.3.7   The normal approximation is $\Phi(\frac{200+0.5-(250,000\times0.0007)}{\sqrt{250,000\times0.0007\times0.9993}}) = 0.9731.$

5.3.8   (a) The normal approximation is $1 - \Phi(\frac{30-0.5-(60\times0.25)}{\sqrt{60\times0.25\times0.75}}) \simeq 0.$

      (b) It is required that $P(B(n, 0.25) \leq 0.35n) \geq 0.99$ which using the normal approximation can be written

$$\Phi\left(\frac{0.35n + 0.5 - 0.25n}{\sqrt{n \times 0.25 \times 0.75}}\right) \geq 0.99.$$

      This is satisfied for $n \geq 92$.

5.3.9   The yearly take can be approximated by a normal distribution with $\mu = 365 \times \frac{5}{0.9} = 2,027.8$ and $\sigma^2 = 365 \times \frac{5}{0.9^2} = 2,253.1$.
$P(N(2027.8, 2253.1) \geq 2,000) = 0.7210.$

5.4 Distributions Related to the Normal Distribution

5.4.1   (a) $E(X) = e^{8.0 + (3.2^2/2)} = 498,800$.

       (b) $Var(X) = e^{(2 \times 8.0) + 3.2^2} \times (e^{3.2^2} - 1) = 6.967 \times 10^{15}$.

       (c)   Since $z_{0.25} = 0.6745$ the upper quartile is $e^{8.0 + (3.2 \times 0.6745)} = 25,807$.

       (d)   The lower quartile is $e^{8.0 + (3.2 \times (-0.6745))} = 344$.

       (e)   The interquartile range is $25,807 - 344 = 25,463$.

       (f) $P(100 \leq X \leq 100,000) = \Phi(\frac{\ln(100,000) - 8.0}{3.2}) - \Phi(\frac{\ln(100) - 8.0}{3.2}) = 0.7195$.

5.4.2   (a) $E(X) = e^{-0.3 + (1.1^2/2)} = 1.357$.

       (b) $Var(X) = e^{(2 \times (-0.3)) + 1.1^2} \times (e^{1.1^2} - 1) = 4.331$.

       (c)   Since $z_{0.25} = 0.6745$ the upper quartile is $e^{-0.3 + (1.1 \times 0.6745)} = 1.556$.

       (d)   The lower quartile is $e^{-0.3 + (1.1 \times (-0.6745))} = 0.353$.

       (e)   The interquartile range is $1.556 - 0.353 = 1.203$.

       (f) $P(0.1 \leq X \leq 7) = \Phi(\frac{\ln(7) - (-0.3)}{1.1}) - \Phi(\frac{\ln(0.1) - (-0.3)}{1.1}) = 0.9451$.

5.4.4   (a) $E(X) = e^{2.3 + (0.2^2/2)} = 10.18$.

       (b)   The median is $e^{2.3} = 9.974$.

       (c)   Since $z_{0.25} = 0.6745$ the upper quartile is $e^{2.3 + (0.2 \times 0.6745)} = 11.41$.

       (d) $P(X \geq 15) = 1 - \Phi(\frac{\ln(15) - 2.3}{0.2}) = 0.0207$.

       (e) $P(X \leq 6) = \Phi(\frac{\ln(6) - 2.3}{0.2}) = 0.0055$.

5.4.5   (a) $\chi^2_{0.10,9} = 14.68$.

       (b) $\chi^2_{0.05,20} = 31.41$.

(c) $\chi^2_{0.01,26} = 45.64$.

(d) $\chi^2_{0.90,50} = 39.69$.

(e) $\chi^2_{0.95,6} = 1.635$.

5.4.6   (a) $\chi^2_{0.12,8} = 12.77$.

(b) $\chi^2_{0.54,19} = 17.74$.

(c) $\chi^2_{0.023,32} = 49.86$.

(d) $P(X \leq 13.3) = 0.6524$.

(e) $P(9.6 \leq X \leq 15.3) = 0.4256$.

5.4.7   (a) $t_{0.10,7} = 1.415$.

(b) $t_{0.05,19} = 1.729$.

(c) $t_{0.01,12} = 2.681$.

(d) $t_{0.025,30} = 2.042$.

(e) $t_{0.005,4} = 4.604$.

5.4.8   (a) $t_{0.27,14} = 0.6282$.

(b) $t_{0.09,22} = 1.385$.

(c) $t_{0.016,7} = 2.670$.

(d) $P(X \leq 1.78) = 0.9556$.

(e) $P(-0.65 \leq X \leq 2.98) = 0.7353$.

(f) $P(|X| \geq 3.02) = 0.0062$.

5.4.9   (a) $F_{0.10,9,10} = 2.347$.

(b) $F_{0.05,6,20} = 2.599$.

(c) $F_{0.01,15,30} = 2.700$.

(d) $F_{0.05,4,8} = 3.838$.

(e) $F_{0.01,20,13} = 3.665$.

5.4.10 (a) $F_{0.04,7,37} = 2.393$.

(b) $F_{0.87,17,43} = 0.6040$.

(c) $F_{0.035,3,8} = 4.732$.

(d) $P(X \geq 2.35) = 0.0625$.

(e) $P(0.21 \leq X \leq 2.92) = 0.9286$.

5.5   Supplementary Problems

5.5.1   (a) $P(N(500, 50^2) \geq 625) = 0.0062$.

      (b) Solving $P(N(500, 50^2) \leq x) = 0.99$ gives $x = 616.3$.

      (c) $P(N(500, 50^2) \geq 700) \simeq 0$.
           Strong suggestion that an eruption is imminent.

5.5.2   (a) $P(N(12500, 200000) \geq 13{,}000) = 0.1318$.

      (b) $P(N(12500, 200000) \leq 11{,}400) = 0.0070$.

      (c) $P(12{,}200 \leq N(12500, 200000) \leq 14{,}000) = 0.7484$.

      (d) Solving $P(N(12500, 200000) \leq x) = 0.95$ gives $x = 13{,}200$.

5.5.3   (a) $P(N(70, 5.4^2) \geq 80) = 0.0320$.

      (b) $P(N(70, 5.4^2) \leq 55) = 0.0027$.

      (c) $P(65 \leq N(70, 5.4^2) \leq 78) = 0.7536$.

      (d) Need $c = \sigma \times z_{0.025} = 5.4 \times 1.9600 = 10.584$.

5.5.4   (a) $P(X_1 - X_2 \geq 0) = P(N(0, 2 \times 5.4^2) \geq 0) = 0.5$.

      (b) $P(X_1 - X_2 \geq 10) = P(N(0, 2 \times 5.4^2) \geq 10) = 0.0952$.

      (c) $P(\frac{X_1 + X_2}{2} - X_3 \geq 10) = P(N(0, 1.5 \times 5.4^2) \geq 10) = 0.0653$.

5.5.5 $P(|X_1 - X_2| \leq 3) = P(|N(0, 2 \times 2^2)| \leq 3) = P(-3 \leq N(0, 8) \leq 3) = 0.7112$.

5.5.6 $E(X) = \frac{1.43 + 1.60}{2} = 1.515$.

      $Var(X) = \frac{(1.60 - 1.43)^2}{12} = 0.002408$.

      The required probability can be estimated as
      $P(170 \leq N(120 \times 1.515, 120 \times 0.002408) \leq 190) \simeq 1$.

5.5.7 $E(X) = \frac{1}{0.31} = 3.2258.$

$Var(X) = \frac{1}{0.31^2} = 10.406.$

The required probability can be estimated as

$P(3.10 \leq N(2000 \times 3.2258, 2000 \times 10.406) \leq 3.25) = 0.5908.$

5.5.8   The required probability is $P(B(350000, 0.06) \geq 20,000).$

The normal approximation is $1 - \Phi(\frac{20,000-0.5-(350,000\times0.06)}{\sqrt{350,000\times0.06\times0.94}}) \simeq 1.$

5.5.9   (a)   The median is $e^{5.5} = 244.7.$

Since $z_{0.25} = 0.6745$ the upper quartile is $e^{5.5+(2.0\times0.6745)} = 942.9.$
The lower quartile is $e^{5.5-(2.0\times0.6745)} = 63.50.$

(b) $P(X \geq 75,000) = 1 - \Phi(\frac{\ln(75,000)-5.5}{2.0}) = 0.0021.$

(c) $P(X \leq 1,000) = \Phi(\frac{\ln(1,000)-5.5}{2.0}) = 0.7592.$

Chapter 6

Descriptive Statistics

6.1 Experimentation

6.1.1   In this case the population is the rather imaginary concept of "all possible die rolls." The sample should be representative assuming that the die is shaken properly.

6.1.2   The population may be all television sets shipped during a certain period of time although the representativeness of the sample depends on whether the television sets shipped on that Friday morning are in any way different from television sets shipped at other times.

6.1.3   Is the population all students? - or the general public? - or perhaps it should just be computing students at that college. You have to consider whether the eye colors of computing students accurately reflect the eye colors of all students or of all people (perhaps eye colors are affected by race and the racial make-up of the class may not reflect that of the student body or the general public as a whole).

6.1.4   The population is all service times under certain conditions. The conditions depend upon how reflective the period between 2:00 and 3:00 on that Saturday afternoon is of other serving periods. The service times may be expected to depend upon how busy the restaurant is and the number of servers available.

6.1.5   The population is all peach boxes received by the supermarket within the time period. The random sampling within each day's shipment and the recording of an observation every day should ensure that the sample is fairly representative.

6.1.6   The population is the number of calls received in each minute of every day during the period of investigation. The spacing of the sample minutes should ensure that the sample is representative.

6.1.7   The population is all bricks shipped by that company. The random selection of the sample should ensure that it is representative.

6.1.8   The population is all car panels spray painted by the machine. The selection of the sample should ensure that it is representative.

6.1.9   The population is all plastic panels made by the machine. If the 80 sample panels are selected in some random manner then they should be representative of the population.

6.2 Data Presentation

6.2.3   The smallest observation 1.097 and the largest observation 1.303 both appear to be outliers.

6.2.4   The largest observation 66.00 can be considered to be an outlier. In addition, the second largest observation 51 might also be considered an outlier.

6.2.5   There would appear to be no reason to doubt that the die is a fair one.

6.2.6   It appears that worse grades are assigned less frequently than better grades.

6.2.7   The assignment "other" is employed considerably less frequently than blue, green, and brown which are each about equally frequent.

6.2.8   The data set appears slightly skewed. The observations 186, 177, 143, and 135 can all be considered to be outliers.

6.2.9   The observations 25 and 14 can be considered to be outliers.

6.2.10   The histogram is bimodal. Possibly it may be considered to be a mix of two distributions corresponding to "busy" periods and "slow" periods.

6.2.11   The smallest observation 0.874 can be considered to be an outlier.

6.2.12   The largest observation 0.538 can be considered to be an outlier.

6.2.13   A skewed data set. The smallest observations 6.00 and 6.04 can be considered to be outliers and possibly some of the other small observations may also be considered to be outliers.

6.3   Sample Statistics

Note: The sample statistics for the problems in this section depend upon whether any observations have been removed as outliers. To avoid confusion, the answers given here assume that **no** observations have been removed.

The trimmed means given here are those obtained by removing the largest 5% and the smallest 5% of the data observations.

6.3.1   The sample mean is $\bar{x} = 155.95$.

The sample median is 159.

The sample trimmed mean is 156.50.

The sample standard deviation is $s = 18.43$.

The upper sample quartile is 169.5.

The lower sample quartile is 143.25.

6.3.2   The sample mean is $\bar{x} = 1.2006$.

The sample median is 1.2010.

The sample trimmed mean is 1.2007.

The sample standard deviation is $s = 0.0291$.

The upper sample quartile is 1.2097.

The lower sample quartile is 1.1890.

6.3.3   The sample mean is $\bar{x} = 37.08$.

The sample median is 35.

The sample trimmed mean is 36.35.

The sample standard deviation is $s = 8.32$.

The upper sample quartile is 40.

The lower sample quartile is 33.5.

6.3.4   The sample mean is $\bar{x} = 3.567$.

The sample median is 3.5.

The sample trimmed mean is 3.574.

The sample standard deviation is $s = 1.767$.

The upper sample quartile is 5.

The lower sample quartile is 2.

6.3.5   The sample mean is $\bar{x} = 69.35$.

The sample median is 66.

The sample trimmed mean is 67.88.

The sample standard deviation is $s = 17.59$.

The upper sample quartile is 76.

The lower sample quartile is 61.

6.3.6   The sample mean is $\bar{x} = 3.291$.

The sample median is 2.

The sample trimmed mean is 2.755.

The sample standard deviation is $s = 3.794$.

The upper sample quartile is 4.

The lower sample quartile is 1.

6.3.7   The sample mean is $\bar{x} = 12.211$.

The sample median is 12.

The sample trimmed mean is 12.175.

The sample standard deviation is $s = 2.629$.

The upper sample quartile is 14.

The lower sample quartile is 10.

6.3.8   The sample mean is $\bar{x} = 1.1106$.

The sample median is 1.1102.

The sample trimmed mean is 1.1112.

The sample standard deviation is $s = 0.0530$.

The upper sample quartile is 1.1400.

The lower sample quartile is 1.0813.

6.3.9   The sample mean is $\bar{x} = 0.23181$.

The sample median is 0.220.

The sample trimmed mean is 0.22875.

The sample standard deviation is $s = 0.07016$.

The upper sample quartile is 0.280.

The lower sample quartile is 0.185.

6.3.10 The sample mean is $\bar{x} = 9.2294$.

The sample median is 9.435.

The sample trimmed mean is 9.3165.

The sample standard deviation is $s = 0.8423$.

The upper sample quartile is 9.81.

The lower sample quartile is 8.9825.

6.5 Supplementary Problems

6.5.1 The population from which the sample is drawn should be all the birds on the island. However, the sample may not be representative if some species are more likely to be observed than others.

It appears that the grey markings are the most common followed by the black markings.

6.5.2 There do not appear to be any seasonal effects although there may possibly be a correlation from one month to the next.

The sample mean is $\bar{x} = 17.79$.

The sample median is 17.

The sample trimmed mean is 17.36.

The sample standard deviation is $s = 6.16$.

The upper sample quartile is 21.75.

The lower sample quartile is 14.

6.5.3 One question of interest in interpreting the data set is whether the month of sampling is representative of other months.

The sample is skewed.

The most frequent data value (the sample mode) is one error.

The sample mean is $\bar{x} = 1.633$.

The sample median is 1.5.

The sample trimmed mean is 1.615.

The sample standard deviation is $s = 0.999$.

The upper sample quartile is 2.

The lower sample quartile is 1.

6.5.4 The population could be all adult males who visit the clinic. This could be representative of all adult males in the population unless there is something special about the clientele of the clinic.

The largest observation 75.9 looks like an outlier on a histogram but may be a valid observation.

The sample mean is $\bar{x} = 69.618$.

The sample median is 69.5.

The sample trimmed mean is 69.513.

The sample standard deviation is $s = 1.523$.

The upper sample quartile is 70.275.

The lower sample quartile is 68.6.

6.5.5   Two or three of the smallest observations and the largest observation may be considered to be outliers.

The sample mean is $\bar{x} = 32.042$.

The sample median is 32.55.

The sample trimmed mean is 32.592.

The sample standard deviation is $s = 5.817$.

The upper sample quartile is 35.5.

The lower sample quartile is 30.425.

Chapter 7

Statistical Estimation and Sampling Distributions

7.2 Properties of Point Estimates

7.2.1   (a) $\text{bias}(\hat{\mu}_1) = 0$. The point estimate $\hat{\mu}_1$ is unbiased.

                $\text{bias}(\hat{\mu}_2) = 0$. The point estimate $\hat{\mu}_2$ is unbiased.

                $\text{bias}(\hat{\mu}_3) = 9 - \frac{\mu}{2}$.

      (b) $Var(\hat{\mu}_1) = 6.25$.

                $Var(\hat{\mu}_2) = 9.0625$.

                $Var(\hat{\mu}_3) = 1.9444$. The point estimate $\hat{\mu}_3$ has the smallest variance.

      (c) $MSE(\hat{\mu}_1) = 6.25$.

                $MSE(\hat{\mu}_2) = 9.0625$.

                $MSE(\hat{\mu}_3) = 1.9444 + (9 - \frac{\mu}{2})^2$. This is equal to 26.9444 when $\mu = 8$.

7.2.2   (a) $\text{bias}(\hat{\mu}_1) = 0$. The point estimate $\hat{\mu}_1$ is unbiased.

                $\text{bias}(\hat{\mu}_2) = -0.217\mu$.

                $\text{bias}(\hat{\mu}_3) = 2 - \frac{\mu}{4}$.

      (b) $Var(\hat{\mu}_1) = 4.444$.

                $Var(\hat{\mu}_2) = 2.682$. The point estimate $\hat{\mu}_2$ has the smallest variance.

                $Var(\hat{\mu}_3) = 2.889$.

      (c) $MSE(\hat{\mu}_1) = 4.444$.

                $MSE(\hat{\mu}_2) = 2.682 + 0.0469\mu^2$. This is equal to 3.104 when $\mu = 3$.

                $MSE(\hat{\mu}_3) = 2.889 + (2 - \frac{\mu}{4})^2$. This is equal to 4.452 when $\mu = 3$.

7.2.3   (a) $Var(\hat{\mu}_1) = 2.5$.

(b) The value $p = 0.6$ produces the smallest variance which is $Var(\hat{\mu}) = 2.4$.

(c) The relative efficiency is $\frac{2.4}{2.5} = 0.96$.

7.2.4   (a) $Var(\hat{\mu}_1) = 2$.

(b) The value $p = 0.875$ produces the smallest variance which is $Var(\hat{\mu}) = 0.875$.

(c) The relative efficiency is $\frac{0.875}{2} = 0.4375$.

7.2.5   (a) $a_1 + \ldots + a_n = 1$.

(b) $a_1 = \ldots = a_n = \frac{1}{n}$.

7.2.6 $MSE(\hat{\theta}_1) = 0.02\, \theta^2 + (0.13\,\theta)^2 = 0.0369\,\theta^2$.
$MSE(\hat{\theta}_2) = 0.07\, \theta^2 + (0.05\,\theta)^2 = 0.0725\,\theta^2$.
$MSE(\hat{\theta}_3) = 0.005\, \theta^2 + (0.24\,\theta)^2 = 0.0626\,\theta^2$.
The point estimate $\hat{\theta}_1$ has the smallest mean square error.

7.2.7 $bias(\hat{\mu}) = \frac{\mu_0 - \mu}{2}$.
$Var(\hat{\mu}) = \frac{\sigma^2}{4}$.
$MSE(\hat{\mu}) = \frac{\sigma^2}{4} + \frac{(\mu_0 - \mu)^2}{4}$.
$MSE(X) = \sigma^2$.

7.2.8   (a) $bias(\hat{p}) = -\frac{p}{11}$.

(b) $Var(\hat{p}) = \frac{10\,p\,(1-p)}{121}$.

(c) $MSE(\hat{p}) = \frac{10\,p\,(1-p)}{121} + \left(\frac{p}{11}\right)^2 = \frac{10p - 9p^2}{121}$.

(d) $bias(X/10) = 0$.
$Var(X/10) = \frac{p(1-p)}{10}$.
$MSE(X/10) = \frac{p(1-p)}{10}$.

7.3 Sampling Distributions

7.3.1 $Var(X_1/n_1) = \frac{p(1-p)}{n_1}$.

$Var(X_2/n_2) = \frac{p(1-p)}{n_2}$.

The relative efficiency is the ratio of these two variances which is $\frac{n_1}{n_2}$. $\frac{n_2}{n_1}$

7.3.2 (a) $P(|N(0,\frac{1}{10})| \le 0.3) = 0.6572$.

(b) $P(|N(0,\frac{1}{30})| \le 0.3) = 0.8996$.

7.3.3 (a) $P(|N(0,\frac{7}{15})| \le 0.4) = 0.8786$.

(b) $P(|N(0,\frac{7}{50})| \le 0.4) = 0.9954$.

7.3.4 (a) Solving $P(5\,\chi^2_{30}/30 \le c) = P(\chi^2_{30} \le 6c) = 0.90$ gives $c = 6.709$.

(b) Solving $P(5\,\chi^2_{30}/30 \le c) = P(\chi^2_{30} \le 6c) = 0.95$ gives $c = 7.296$.

7.3.5 (a) Solving $P(32\,\chi^2_{20}/20 \le c) = P(\chi^2_{20} \le 5c/8) = 0.90$ gives $c = 45.46$.

(b) Solving $P(32\,\chi^2_{20}/20 \le c) = P(\chi^2_{20} \le 5c/8) = 0.95$ gives $c = 50.26$.

7.3.6 (a) Solving $P(|t_{15}| \le c) = 0.95$ gives $c = t_{0.025,15} = 2.131$.

(b) Solving $P(|t_{15}| \le c) = 0.99$ gives $c = t_{0.005,15} = 2.947$.

7.3.7 (a) Solving $P(|t_{20}/\sqrt{21}| \le c) = 0.95$ gives $c = t_{0.025,20}/\sqrt{21} = 0.4552$.

(b) Solving $P(|t_{20}/\sqrt{21}| \le c) = 0.99$ gives $c = t_{0.005,20}/\sqrt{21} = 0.6209$.

7.3.8 $\hat{p} = \frac{234}{450} = 0.52$.

$$s.e.(\hat{p}) = \sqrt{\frac{\hat{p}\,(1-\hat{p})}{n}} = \sqrt{\frac{0.52 \times 0.48}{450}} = 0.0236.$$

7.3.9 $\hat{\mu} = \bar{x} = 974.3$.

$$s.e.(\bar{x}) = \sqrt{\frac{452.1}{35}} = 3.594.$$

7.3.10 $\hat{p} = \frac{24}{120} = 0.2$.

$$s.e.(\hat{p}) = \sqrt{\frac{\hat{p}(1-\hat{p})}{n}} = \sqrt{\frac{0.2 \times 0.8}{120}} = 0.0365.$$

7.3.11 $\hat{p} = \frac{33}{150} = 0.22$.

$$s.e.(\hat{p}) = \sqrt{\frac{\hat{p}(1-\hat{p})}{n}} = \sqrt{\frac{0.22 \times 0.78}{150}} = 0.0338.$$

7.3.12 $\hat{p} = \frac{26}{80} = 0.325$.

$$s.e.(\hat{p}) = \sqrt{\frac{\hat{p}(1-\hat{p})}{n}} = \sqrt{\frac{0.325 \times 0.675}{80}} = 0.0524.$$

7.3.13 $\hat{\mu} = \bar{x} = 69.35$.

$$s.e.(\bar{x}) = \frac{s}{\sqrt{n}} = \frac{17.59}{\sqrt{200}} = 1.244.$$

7.3.14 $\hat{\mu} = \bar{x} = 3.291$.

$$s.e.(\bar{x}) = \frac{s}{\sqrt{n}} = \frac{3.794}{\sqrt{55}} = 0.512.$$

7.3.15 $\hat{\mu} = \bar{x} = 12.211$.

$$s.e.(\bar{x}) = \frac{s}{\sqrt{n}} = \frac{2.629}{\sqrt{90}} = 0.277.$$

7.3.16 $\hat{\mu} = \bar{x} = 1.1106$.

$$s.e.(\bar{x}) \; = \; \frac{s}{\sqrt{n}} \; = \; \frac{0.0530}{\sqrt{125}} \; = \; 0.00474.$$

7.3.17 $\hat{\mu} = \bar{x} = 0.23181$.

$$s.e.(\bar{x}) \; = \; \frac{s}{\sqrt{n}} \; = \; \frac{0.07016}{\sqrt{75}} \; = \; 0.00810.$$

7.3.18 $\hat{\mu} = \bar{x} = 9.2294$.

$$s.e.(\bar{x}) \; = \; \frac{s}{\sqrt{n}} \; = \; \frac{0.8423}{\sqrt{80}} \; = \; 0.0942.$$

7.4   Constructing Parameter Estimates

7.4.1 $\hat{\lambda} = \bar{x} = 5.63$.

$$s.e.(\hat{\lambda}) = \sqrt{\frac{\hat{\lambda}}{n}} = \sqrt{\frac{5.63}{23}} = 0.495.$$

7.4.2   Using the method of moments the point estimates $\hat{a}$ and $\hat{b}$ are the solutions to the equations

$$\frac{a}{a+b} = 0.782$$

and

$$\frac{ab}{(a+b)^2(a+b+1)} = 0.0083,$$

which are $\hat{a} = 15.28$ and $\hat{b} = 4.26$.

7.4.3   Using the method of moments

$$E(X) = \frac{1}{\lambda} = \bar{x}$$

which gives $\hat{\lambda} = \frac{1}{\bar{x}}$.

The likelihood is

$$L(x_1, \ldots, x_n, \lambda) = \lambda^n \, e^{-\lambda(x_1 + \ldots + x_n)}$$

which is maximized at $\hat{\lambda} = \frac{1}{\bar{x}}$.

7.4.4 $\hat{p}_i = \frac{x_i}{n}$ for $1 \leq i \leq n$.

7.4.5   Using the method of moments

$$E(X) = \frac{5}{\lambda} = \bar{x}$$

which gives $\hat{\lambda} = \frac{5}{\bar{x}}$.

The likelihood is

$$L(x_1, \ldots, x_n, \lambda) = \left(\frac{1}{24}\right)^n \times \lambda^{5n} \times x_1^4 \times \ldots \times x_n^4 \times e^{-\lambda(x_1 + \ldots + x_n)}$$

which is maximimized at $\hat{\lambda} = \frac{5}{\bar{x}}$.

7.5 Supplementary Problems

7.5.1 $\text{bias}(\hat{\mu}_1) = 5 - \frac{\mu}{2}$.

$\text{bias}(\hat{\mu}_2) = 0$.

$Var(\hat{\mu}_1) = \frac{1}{8}$.
$Var(\hat{\mu}_2) = \frac{1}{2}$.

$MSE(\hat{\mu}_1) = \frac{1}{8} + (5 - \frac{\mu}{2})^2$.
$MSE(\hat{\mu}_2) = \frac{1}{2}$.

7.5.2 (a) $\text{bias}(\hat{p}) = -\frac{p}{7}$.

(b) $Var(\hat{p}) = \frac{3p(1-p)}{49}$.

(c) $MSE(\hat{p}) = \frac{3p(1-p)}{49} + (\frac{p}{7})^2 = \frac{3p-2p^2}{49}$.

(d) $MSE(X/12) = \frac{p(1-p)}{12}$.

7.5.3 (a) $F(t) = P(T \le t) = P(X_1 \le t) \times \ldots \times P(X_n \le t)$
$= \frac{t}{\theta} \times \ldots \times \frac{t}{\theta} = (\frac{t}{\theta})^n$ for $0 \le t \le \theta$.

(b) $f(t) = \frac{dF(t)}{dt} = n\frac{t^{n-1}}{\theta^n}$ for $0 \le t \le \theta$.

(c) Notice that

$$E(T) = \int_0^\theta t\, f(t)\, dt = \frac{n}{n+1}\theta,$$

so that $E(\hat{\theta}) = \theta$.

(d) Notice that

$$E(T^2) = \int_0^\theta t^2\, f(t)\, dt = \frac{n}{n+2}\theta^2,$$

so that

$$Var(T) = \frac{n}{n+2}\theta^2 - \left(\frac{n}{n+1}\theta\right)^2 = \frac{n\theta^2}{(n+2)(n+1)^2}.$$

Consequently,

$$Var(\hat{\theta}) = \frac{(n+1)^2}{n^2}Var(T) = \frac{\theta^2}{n(n+2)}$$

and

$$s.e.(\hat{\theta}) \ = \ \frac{\hat{\theta}}{\sqrt{n(n+2)}}.$$

(e) $\hat{\theta} = \frac{11}{10} \times 7.3 = 8.03.$
$s.e.(\hat{\theta}) = \frac{8.03}{\sqrt{10 \times 12}} = 0.733.$

7.5.4 Recall that $f(x_i, \theta) = \frac{1}{\theta}$ for $0 \le x_i \le \theta$ (and $f(x_i, \theta) = 0$ elsewhere) so that the likelihood is $\frac{1}{\theta^n}$ as long as $x_i \le \theta$ for $1 \le i \le n$ and is equal to zero otherwise.
$\text{bias}(\hat{\theta}) = -\frac{\theta}{n+1}.$

7.5.5 Using the method of moments

$$E(X) \ = \ \frac{1}{p} \ = \ \bar{x}$$

which gives $\hat{p} = \frac{1}{\bar{x}}$.
The likelihood is

$$L(x_1, \ldots, x_n, \lambda) \ = \ p^n (1-p)^{x_1 + \ldots + x_n - n}$$

which is maximimized at $\hat{p} = \frac{1}{\bar{x}}$.

7.5.6 $\hat{p} = \frac{35}{100} = 0.35.$

$$s.e.(\hat{p}) \ = \ \sqrt{\frac{\hat{p}\,(1-\hat{p})}{n}} \ = \ \sqrt{\frac{0.35 \times 0.65}{100}} \ = \ 0.0477.$$

7.5.7 $\hat{\mu} = \bar{x} = 17.79.$

$$s.e.(\bar{x}) \ = \ \frac{s}{\sqrt{n}} \ = \ \frac{6.16}{\sqrt{24}} \ = \ 1.26.$$

7.5.8 $\hat{\mu} = \bar{x} = 1.633.$

$$s.e.(\bar{x}) \ = \ \frac{s}{\sqrt{n}} \ = \ \frac{0.999}{\sqrt{30}} \ = \ 0.182.$$

7.5.9 $\hat{\mu} = \bar{x} = 69.618.$

$$s.e.(\bar{x}) \; = \; \frac{s}{\sqrt{n}} \; = \; \frac{1.523}{\sqrt{60}} \; = \; 0.197.$$

7.5.10 $\hat{\mu} = \bar{x} = 32.042.$

$$s.e.(\bar{x}) \; = \; \frac{s}{\sqrt{n}} \; = \; \frac{5.817}{\sqrt{40}} \; = \; 0.920.$$

Chapter 8

Inferences on a Population Mean

8.1 Confidence Intervals

8.1.1   With $t_{0.025,30} = 2.042$ the confidence interval is

$(53.42 - \frac{2.042 \times 3.05}{\sqrt{31}}, 53.42 + \frac{2.042 \times 3.05}{\sqrt{31}}) = (52.30, 54.54)$.

8.1.2   With $t_{0.005,40} = 2.704$ the confidence interval is

$(3.04 - \frac{2.704 \times 0.124}{\sqrt{41}}, 3.04 + \frac{2.704 \times 0.124}{\sqrt{41}}) = (2.99, 3.09)$.

The confidence interval does not contain the value 2.90 and so it is not a plausible value for the mean glass thickness.

8.1.3   At 90% confidence $t_{0.05,19} = 1.729$ and the confidence interval is

$(436.5 - \frac{1.729 \times 11.90}{\sqrt{20}}, 436.5 + \frac{1.729 \times 11.90}{\sqrt{20}}) = (431.9, 441.1)$.

At 95% confidence $t_{0.025,19} = 2.093$ and the confidence interval is

$(436.5 - \frac{2.093 \times 11.90}{\sqrt{20}}, 436.5 + \frac{2.093 \times 11.90}{\sqrt{20}}) = (430.9, 442.1)$.

At 99% confidence $t_{0.005,19} = 2.861$ and the confidence interval is

$(436.5 - \frac{2.861 \times 11.90}{\sqrt{20}}, 436.5 + \frac{2.861 \times 11.90}{\sqrt{20}}) = (428.9, 444.1)$.

Even the 99% confidence level confidence interval does not contain the value 450.0 and so it is not a plausible value for the average breaking strength.

8.1.4   With $t_{0.005,15} = 2.947$ the confidence interval is

$(1.053 - \frac{2.947 \times 0.058}{\sqrt{16}}, 1.053 + \frac{2.947 \times 0.058}{\sqrt{16}}) = (1.010, 1.096)$.

The confidence interval contains the value 1.025 and so it is a plausible value for the average weight.

8.1.5   With $z_{0.025} = 1.960$ the confidence interval is

$$\left(0.0328 - \frac{1.960 \times 0.015}{\sqrt{28}}, 0.0328 + \frac{1.960 \times 0.015}{\sqrt{28}}\right) = (0.0272, 0.0384).$$

8.1.6   At 90% confidence $z_{0.05} = 1.645$ and the confidence interval is

$$\left(19.50 - \frac{1.645 \times 1.0}{\sqrt{10}}, 19.50 + \frac{1.645 \times 1.0}{\sqrt{10}}\right) = (18.98, 20.02).$$

At 95% confidence $z_{0.025} = 1.960$ and the confidence interval is

$$\left(19.50 - \frac{1.960 \times 1.0}{\sqrt{10}}, 19.50 + \frac{1.960 \times 1.0}{\sqrt{10}}\right) = (18.88, 20.12).$$

At 99% confidence $z_{0.005} = 2.576$ and the confidence interval is

$$\left(19.50 - \frac{2.576 \times 1.0}{\sqrt{10}}, 19.50 + \frac{2.576 \times 1.0}{\sqrt{10}}\right) = (18.69, 20.31).$$

Even the 90% confidence level confidence interval contains the value 20.0 and so it is a plausible value for the average resilient modulus.

8.1.7   With $t_{0.025, n-1} \simeq 2.0$ a sufficient sample size can be estimated as

$$n \geq 4 \times \left(\frac{t_{0.025, n-1}\, s}{L_0}\right)^2 = 4 \times \left(\frac{2.0 \times 10.0}{5}\right)^2 = 64.$$

A sample size of about $n = 64$ should be sufficient.

8.1.8   With $t_{0.005, n-1} \simeq 3.0$ a sufficient sample size can be estimated as

$$n \geq 4 \times \left(\frac{t_{0.005, n-1}\, s}{L_0}\right)^2 = 4 \times \left(\frac{3.0 \times 0.15}{0.2}\right)^2 = 20.25.$$

A sample size slightly larger than 20 should be sufficient.

8.1.9   A total sample size of

$$n \geq 4 \times \left(\frac{t_{0.025, n_1-1}\, s}{L_0}\right)^2 = 4 \times \left(\frac{2.042 \times 3.05}{2.0}\right)^2 = 38.8$$

is required. Therefore an additional sample of at least $39 - 31 = 8$ observations should be sufficient.

8.1.10   A total sample size of

$$n \geq 4 \times \left(\frac{t_{0.005, n_1-1}\, s}{L_0}\right)^2 = 4 \times \left(\frac{2.704 \times 0.124}{0.05}\right)^2 = 179.9$$

is required. Therefore an additional sample of at least $180 - 41 = 139$ observations should be sufficient.

8.1.11 A total sample size of

$$n \geq 4 \times \left(\frac{t_{0.005,n_1-1} \, s}{L_0} \right)^2 = 4 \times \left(\frac{2.861 \times 11.90}{10.0} \right)^2 = 46.4$$

is required. Therefore an additional sample of at least $47 - 20 = 27$ observations should be sufficient.

8.1.12 With $t_{0.05,29} = 1.699$ the value of c is obtained as

$$c = \bar{x} + \frac{t_{\alpha,n-1} \, s}{\sqrt{n}} = 14.62 + \frac{1.699 \times 2.98}{\sqrt{30}} = 15.54.$$

The confidence interval does not contain the value 16.0 and so it is not plausible that $\mu \geq 16$.

8.1.13 With $t_{0.01,60} = 2.390$ the value of c is obtained as

$$c = \bar{x} - \frac{t_{\alpha,n-1} \, s}{\sqrt{n}} = 0.768 - \frac{2.390 \times 0.0231}{\sqrt{61}} = 0.761.$$

The confidence interval contains the value 0.765 and so it is plausible that the average solution density is less than 0.765.

8.1.14 With $z_{0.05} = 1.645$ the value of c is obtained as

$$c = \bar{x} - \frac{z_\alpha \, \sigma}{\sqrt{n}} = 11.80 - \frac{1.645 \times 2.0}{\sqrt{19}} = 11.05.$$

8.1.15 With $z_{0.01} = 2.326$ the value of c is obtained as

$$c = \bar{x} + \frac{z_\alpha \, \sigma}{\sqrt{n}} = 415.7 + \frac{2.326 \times 10.0}{\sqrt{29}} = 420.0.$$

The confidence interval contains the value 418.0 and so it is plausible that the mean radiation level is greater than 418.0.

Note: The sample statistics for the following problems in this section and the related problems in this chapter depend upon whether any observations have been removed as outliers. To avoid confusion, the answers given here assume that **no** observations have been removed. Notice that removing observations as outliers reduces the sample standard deviation s as well as affecting the sample mean $\bar{x}$.

8.1.16   At 95% confidence $t_{0.025,199} = 1.972$ and the confidence interval is
$$(69.35 - \frac{1.972 \times 17.59}{\sqrt{200}}, 69.35 + \frac{1.972 \times 17.59}{\sqrt{200}}) = (66.89, 71.80).$$

8.1.17   At 95% confidence $t_{0.025,99} = 1.987$ and the confidence interval is
$$(12.211 - \frac{1.987 \times 2.629}{\sqrt{90}}, 12.211 + \frac{1.987 \times 2.629}{\sqrt{90}}) = (11.66, 12.76).$$

8.1.18   At 95% confidence $t_{0.025,124} = 1.979$ and the confidence interval is
$$(1.11059 - \frac{1.979 \times 0.05298}{\sqrt{125}}, 1.11059 + \frac{1.979 \times 0.05298}{\sqrt{125}}) = (1.101, 1.120).$$

8.1.19   At 95% confidence $t_{0.025,74} = 1.9926$ and the confidence interval is
$$(0.23181 - \frac{1.9926 \times 0.07016}{\sqrt{75}}, 0.23181 + \frac{1.9926 \times 0.07016}{\sqrt{75}}) = (0.2157, 0.2480).$$

8.1.20   At 95% confidence $t_{0.025,79} = 1.9905$ and the confidence interval is
$$(9.2294 - \frac{1.9905 \times 0.0942}{\sqrt{80}}, 9.2294 + \frac{1.9905 \times 0.0942}{\sqrt{80}}) = (9.0419, 9.4169).$$

8.2 Hypothesis Testing

8.2.1 (a) The test statistic is

$$t = \frac{\sqrt{n}(\bar{x} - \mu_0)}{s} = \frac{\sqrt{18} \times (57.74 - 55.0)}{11.2} = 1.04.$$

The p-value is

$$2 \times P(X \geq 1.04) = 0.313$$

where the random variable X has a t distribution with $18 - 1 = 17$ degrees of freedom.

(b) The test statistic is

$$t = \frac{\sqrt{n}(\bar{x} - \mu_0)}{s} = \frac{\sqrt{18} \times (57.74 - 65.0)}{11.2} = -2.75.$$

The p-value is

$$P(X \leq -2.75) = 0.0068$$

where the random variable X has a t distribution with $18 - 1 = 17$ degrees of freedom.

8.2.2 (a) The test statistic is

$$t = \frac{\sqrt{n}(\bar{x} - \mu_0)}{s} = \frac{\sqrt{39} \times (5,532 - 5,680)}{287.8} = -3.21.$$

The p-value is

$$2 \times P(X \geq 3.21) = 0.003$$

where the random variable X has a t distribution with $39 - 1 = 38$ degrees of freedom.

(b) The test statistic is

$$t = \frac{\sqrt{n}(\bar{x} - \mu_0)}{s} = \frac{\sqrt{39} \times (5,532 - 5,450)}{287.8} = 1.78.$$

The p-value is

$$P(X \geq 1.78) = 0.042$$

where the random variable X has a t distribution with $39 - 1 = 38$ degrees of freedom.

8.2.3 (a) The test statistic is

$$z = \frac{\sqrt{n}(\bar{x} - \mu_0)}{\sigma} = \frac{\sqrt{13} \times (2.879 - 3.0)}{0.325} = -1.34.$$

The p-value is

$$2 \times \Phi(-1.34) \ = \ 0.180.$$

(b) The test statistic is

$$z \ = \ \frac{\sqrt{n}(\bar{x} - \mu_0)}{\sigma} \ = \ \frac{\sqrt{13} \times (2.879 - 3.1)}{0.325} \ = \ -2.45.$$

The p-value is

$$\Phi(-2.45) \ = \ 0.007.$$

8.2.4 (a) The test statistic is

$$z \ = \ \frac{\sqrt{n}(\bar{x} - \mu_0)}{\sigma} \ = \ \frac{\sqrt{44} \times (87.90 - 90.0)}{5.90} \ = \ -2.36.$$

The p-value is

$$2 \times \Phi(-2.36) \ = \ 0.018.$$

(b) The test statistic is

$$z \ = \ \frac{\sqrt{n}(\bar{x} - \mu_0)}{\sigma} \ = \ \frac{\sqrt{44} \times (87.90 - 86.0)}{5.90} \ = \ 2.14.$$

The p-value is

$$1 - \Phi(2.14) \ = \ 0.016.$$

8.2.5 (a) The critical point is $t_{0.05,40} = 1.684$ and the null hypothesis is accepted when $|t| \le 1.684$.

(b) The critical point is $t_{0.005,40} = 2.704$ and the null hypothesis is rejected when $|t| > 2.704$.

(c) The test statistic is

$$t \ = \ \frac{\sqrt{n}(\bar{x} - \mu_0)}{s} \ = \ \frac{\sqrt{41} \times (3.04 - 3.00)}{0.124} \ = \ 2.066.$$

The null hypothesis is rejected at size $\alpha = 0.10$ and accepted at size $\alpha = 0.01$.

(d) The p-value is

$$2 \times P(X \ge 2.066) \ = \ 0.045$$

where the random variable X has a t distribution with $41 - 1 = 40$ degrees of freedom.

8.2.6   (a) The critical point is $t_{0.05,19} = 1.729$ and the null hypothesis is accepted when $|t| \le 1.729$.

       (b) The critical point is $t_{0.005,19} = 2.861$ and the null hypothesis is rejected when $|t| > 2.861$.

       (c) The test statistic is

$$t = \frac{\sqrt{n}(\bar{x} - \mu_0)}{s} = \frac{\sqrt{20} \times (436.5 - 430.0)}{11.90} = 2.443.$$

The null hypothesis is rejected at size $\alpha = 0.10$ and accepted at size $\alpha = 0.01$.

       (d) The p-value is

$$2 \times P(X \ge 2.443) = 0.025$$

where the random variable X has a t distribution with $20 - 1 = 19$ degrees of freedom.

8.2.7   (a) The critical point is $t_{0.05,15} = 1.753$ and the null hypothesis is accepted when $|t| \le 1.753$.

       (b) The critical point is $t_{0.005,15} = 2.947$ and the null hypothesis is rejected when $|t| > 2.947$.

       (c) The test statistic is

$$t = \frac{\sqrt{n}(\bar{x} - \mu_0)}{s} = \frac{\sqrt{16} \times (1.053 - 1.025)}{0.058} = 1.931.$$

The null hypothesis is rejected at size $\alpha = 0.10$ and accepted at size $\alpha = 0.01$.

       (d) The p-value is

$$2 \times P(X \ge 1.931) = 0.073$$

where the random variable X has a t distribution with $16 - 1 = 15$ degrees of freedom.

8.2.8   (a) The critical point is $z_{0.05} = 1.645$ and the null hypothesis is accepted when $|z| \le 1.645$.

       (b) The critical point is $z_{0.005} = 2.576$ and the null hypothesis is rejected when $|z| > 2.576$.

(c) The test statistic is

$$z = \frac{\sqrt{n}(\bar{x} - \mu_0)}{\sigma} = \frac{\sqrt{10} \times (19.50 - 20.0)}{1.0} = -1.581.$$

The null hypothesis is accepted at size $\alpha = 0.10$ and consequently also at size $\alpha = 0.01$.

(d) The p-value is

$$2 \times \Phi(-1.581) = 0.114.$$

8.2.9 (a) The critical point is $t_{0.10,60} = 1.296$ and the null hypothesis is accepted when $t \leq 1.296$.

(b) The critical point is $t_{0.01,60} = 2.390$ and the null hypothesis is rejected when $t > 2.390$.

(c) The test statistic is

$$t = \frac{\sqrt{n}(\bar{x} - \mu_0)}{s} = \frac{\sqrt{61} \times (0.0768 - 0.065)}{0.0231} = 3.990.$$

The null hypothesis is rejected at size $\alpha = 0.01$ and consequently also at size $\alpha = 0.10$.

(d) The p-value is

$$P(X \geq 3.990) = 0.0001$$

where the random variable X has a t distribution with $61 - 1 = 60$ degrees of freedom.

8.2.10 (a) The critical point is $z_{0.10} = 1.282$ and the null hypothesis is accepted when $z \geq -1.282$.

(b) The critical point is $z_{0.01} = 2.326$ and the null hypothesis is rejected when $z < -2.326$.

(c) The test statistic is

$$z = \frac{\sqrt{n}(\bar{x} - \mu_0)}{\sigma} = \frac{\sqrt{29} \times (415.7 - 420.0)}{10.0} = -2.316.$$

The null hypothesis is rejected at size $\alpha = 0.10$ and accepted at size $\alpha = 0.01$.

(d) The p-value is

$$\Phi(-2.316) = 0.0103.$$

8.2.11 Consider the hypothesis testing problem

$$H_0 : \mu = 44.350 \quad \text{versus} \quad H_A : \mu \neq 44.350.$$

The test statistic is

$$t = \frac{\sqrt{n}(\bar{x} - \mu_0)}{s} = \frac{\sqrt{24} \times (44.364 - 44.350)}{0.019} = 3.61.$$

The p-value is

$$2 \times P(X \geq 3.61) = 0.0014$$

where the random variable X has a t distribution with $24 - 1 = 23$ degrees of freedom. There is sufficient evidence to conclude that the machine is miscalibrated.

8.2.12 Consider the hypothesis testing problem

$$H_0 : \mu \leq 120 \quad \text{versus} \quad H_A : \mu > 120.$$

The test statistic is

$$t = \frac{\sqrt{n}(\bar{x} - \mu_0)}{s} = \frac{\sqrt{36} \times (122.5 - 120.0)}{13.4} = 1.12.$$

The p-value is

$$P(X \geq 1.12) = 0.135$$

where the random variable X has a t distribution with $36 - 1 = 35$ degrees of freedom. There is *not* sufficient evidence to conclude that the manufacturer's claim is incorrect.

8.2.13 Consider the hypothesis testing problem

$$H_0 : \mu \leq 12.50 \quad \text{versus} \quad H_A : \mu > 12.50.$$

The test statistic is

$$t = \frac{\sqrt{n}(\bar{x} - \mu_0)}{s} = \frac{\sqrt{15} \times (14.82 - 12.50)}{2.91} = 3.09.$$

The p-value is

$$P(X \geq 3.09) = 0.004$$

where the random variable X has a t distribution with $15 - 1 = 14$ degrees of freedom. There is sufficient evidence to conclude that the chemical plant is in violation of the working code.

8.2.14   Consider the hypothesis testing problem

$$H_0 : \mu \geq 0.25 \quad \text{versus} \quad H_A : \mu < 0.25.$$

The test statistic is

$$t = \frac{\sqrt{n}(\bar{x} - \mu_0)}{s} = \frac{\sqrt{23} \times (0.228 - 0.250)}{0.0872} = -1.21.$$

The p-value is

$$P(X \leq -1.21) = 0.120$$

where the random variable X has a t distribution with $23 - 1 = 22$ degrees of freedom. There is *not* sufficient evidence to conclude that the advertised claim is false.

8.2.15   Consider the hypothesis testing problem

$$H_0 : \mu \leq 65 \quad \text{versus} \quad H_A : \mu > 65.$$

The test statistic is

$$t = \frac{\sqrt{n}(\bar{x} - \mu_0)}{s} = \frac{\sqrt{200} \times (69.35 - 65.00)}{17.59} = 3.50.$$

The p-value is

$$P(X \geq 3.50) = 0.0003$$

where the random variable X has a t distribution with $200 - 1 = 199$ degrees of freedom.

There is sufficient evidence to conclude that the average service time is greater than 65 seconds and that the manager's claim is incorrect.

8.2.16   Consider the hypothesis testing problem

$$H_0 : \mu \geq 13 \quad \text{versus} \quad H_A : \mu < 13.$$

The test statistic is

$$t = \frac{\sqrt{n}(\bar{x} - \mu_0)}{s} = \frac{\sqrt{90} \times (12.211 - 13.000)}{2.629} = -2.85.$$

The p-value is

$$P(X \leq -2.85) = 0.0027$$

where the random variable X has a t distribution with $90 - 1 = 89$ degrees of freedom.

There is sufficient evidence to conclude that the average number of calls taken per minute is less than 13 so that the manager's claim is false.

8.2.17   Consider the hypothesis testing problem

$$H_0 : \mu = 1.1 \quad \text{versus} \quad H_A : \mu \neq 1.1.$$

The test statistic is

$$t = \frac{\sqrt{n}(\bar{x} - \mu_0)}{s} = \frac{\sqrt{125} \times (1.11059 - 1.10000)}{0.05298} = 2.23.$$

The p-value is

$$2 \times P(X \geq 2.23) = 0.028$$

where the random variable X has a t distribution with $125 - 1 = 124$ degrees of freedom.

There is some evidence that the manufacturing process needs adjusting but it is not overwhelming.

8.2.18   Consider the hypothesis testing problem

$$H_0 : \mu = 0.2 \quad \text{versus} \quad H_A : \mu \neq 0.2.$$

The test statistic is

$$t = \frac{\sqrt{n}(\bar{x} - \mu_0)}{s} = \frac{\sqrt{75} \times (0.23181 - 0.20000)}{0.07016} = 3.93.$$

The p-value is

$$2 \times P(X \geq 3.93) = 0.0002$$

where the random variable X has a t distribution with $75 - 1 = 74$ degrees of freedom.

There is sufficient evidence to conclude that the spray painting machine is not operating properly.

8.2.19   Consider the hypothesis testing problem

$$H_0 : \mu \geq 9.5 \quad \text{versus} \quad H_A : \mu < 9.5.$$

The test statistic is

$$t = \frac{\sqrt{n}(\bar{x} - \mu_0)}{s} = \frac{\sqrt{80} \times (9.2294 - 9.5000)}{0.8423} = -2.87.$$

The p-value is

$$P(X \leq -2.87) = 0.0026$$

where the random variable X has a t distribution with $80 - 1 = 79$ degrees of freedom.

There is sufficient evidence to conclude that the design criterion has not been met.

8.4   Supplementary Problems

8.4.1   (a) Consider the hypothesis testing problem

$$H_0 : \mu \leq 65 \quad \text{versus} \quad H_A : \mu > 65.$$

The test statistic is

$$t = \frac{\sqrt{n}(\bar{x} - \mu_0)}{s} = \frac{\sqrt{15} \times (67.42 - 65.00)}{4.947} = 1.89.$$

The p-value is

$$P(X \geq 1.89) = 0.040$$

where the random variable X has a t distribution with $15 - 1 = 14$ degrees of freedom.

There is some evidence that the average distance at which the target is detected is at least 65 miles although the evidence is not overwhelming.

(b) With $t_{0.01,14} = 2.624$ the confidence interval is
$(67.42 - \frac{2.624 \times 4.947}{\sqrt{15}}, \infty) = (64.07, \infty).$

8.4.2   (a) Consider the hypothesis testing problem

$$H_0 : \mu \geq 10 \quad \text{versus} \quad H_A : \mu < 10.$$

The test statistic is

$$t = \frac{\sqrt{n}(\bar{x} - \mu_0)}{s} = \frac{\sqrt{40} \times (9.39 - 10.00)}{1.041} = -3.71.$$

The p-value is

$$P(X \leq -3.71) = 0.0003$$

where the random variable X has a t distribution with $40 - 1 = 39$ degrees of freedom.

The company can safely conclude that the telephone surveys will last on average less than ten minutes each.

(b) With $t_{0.01,39} = 2.426$ the confidence interval is
$(-\infty, 9.39 + \frac{2.426 \times 1.041}{\sqrt{40}}) = (-\infty, 9.79).$

8.4.3   (a) Consider the hypothesis testing problem

$$H_0 : \mu = 75.0 \quad \text{versus} \quad H_A : \mu \neq 75.0.$$

The test statistic is

$$t = \frac{\sqrt{n}(\bar{x} - \mu_0)}{s} = \frac{\sqrt{30} \times (74.43 - 75.00)}{2.095} = -1.49.$$

The p-value is

$$2 \times P(X \le -1.49) = 0.147$$

where the random variable X has a t distribution with $30 - 1 = 29$ degrees of freedom.

There is not sufficient evidence to conclude that the paper does not have an average weight of 75.0 g/m^2.

(b) With $t_{0.005,29} = 2.756$ the confidence interval is
$$\left(74.43 - \frac{2.756 \times 2.095}{\sqrt{30}}, 74.43 + \frac{2.756 \times 2.095}{\sqrt{30}}\right) = (73.38, 75.48).$$

(c) A total sample size of

$$n \ge 4 \times \left(\frac{t_{0.005,n_1-1} \, s}{L_0}\right)^2 = 4 \times \left(\frac{2.756 \times 2.095}{1.5}\right)^2 = 59.3$$

is required. Therefore an additional sample of at least $60 - 30 = 30$ observations should be sufficient.

8.4.4   (a) Consider the hypothesis testing problem

$$H_0 : \mu \ge 0.50 \quad \text{versus} \quad H_A : \mu < 0.50.$$

The test statistic is

$$t = \frac{\sqrt{n}(\bar{x} - \mu_0)}{s} = \frac{\sqrt{14} \times (0.497 - 0.500)}{0.0764} = -0.147.$$

The p-value is

$$P(X \le -0.147) = 0.443$$

where the random variable X has a t distribution with $14 - 1 = 13$ degrees of freedom.

There is not sufficient evidence to establish that the average deformity value of diseased arteries is less than 0.50.

(b) With $t_{0.005,13} = 3.012$ the confidence interval is
$$\left(0.497 - \frac{3.012 \times 0.0764}{\sqrt{14}}, 0.497 + \frac{3.012 \times 0.0764}{\sqrt{14}}\right) = (0.435, 0.559).$$

(c) A total sample size of

$$n \ge 4 \times \left(\frac{t_{0.005,n_1-1} \, s}{L_0}\right)^2 = 4 \times \left(\frac{3.012 \times 0.0764}{0.10}\right)^2 = 21.2$$

is required. Therefore an additional sample of at least $22 - 14 = 8$ observations should be sufficient.

8.4.5   At 90% confidence $t_{0.05,59} = 1.671$ and the confidence interval is
$(69.618 - \frac{1.671 \times 1.523}{\sqrt{60}}, 69.618 + \frac{1.671 \times 1.523}{\sqrt{60}}) = (69.29, 69.95)$.

At 95% confidence $t_{0.025,59} = 2.001$ and the confidence interval is
$(69.618 - \frac{2.001 \times 1.523}{\sqrt{60}}, 69.618 + \frac{2.001 \times 1.523}{\sqrt{60}}) = (69.23, 70.01)$.

At 99% confidence $t_{0.005,59} = 2.662$ and the confidence interval is
$(69.618 - \frac{2.662 \times 1.523}{\sqrt{60}}, 69.618 + \frac{2.662 \times 1.523}{\sqrt{60}}) = (69.10, 70.14)$.

There is not strong evidence that 70 inches is not a plausible value for the mean height because it is included in the 95% confidence level confidence interval.

8.4.6   At 90% confidence $t_{0.05,39} = 1.685$ and the confidence interval is
$(32.042 - \frac{1.685 \times 5.817}{\sqrt{40}}, 32.042 + \frac{1.685 \times 5.817}{\sqrt{40}}) = (30.49, 33.59)$.

At 95% confidence $t_{0.025,39} = 2.023$ and the confidence interval is
$(32.042 - \frac{2.023 \times 5.817}{\sqrt{40}}, 32.042 + \frac{2.023 \times 5.817}{\sqrt{40}}) = (30.18, 33.90)$.

At 99% confidence $t_{0.005,39} = 2.708$ and the confidence interval is
$(32.042 - \frac{2.708 \times 5.817}{\sqrt{40}}, 32.042 + \frac{2.708 \times 5.817}{\sqrt{40}}) = (29.55, 34.53)$.

Since 35 and larger values are not contained within the 99% confidence level confidence interval they are not plausible values for the mean shoot height and so these results contradict the results of the previous study.

8.4.8   At 95% confidence $\chi^2_{0.025,17} = 30.19$ and $\chi^2_{0.975,17} = 7.564$ so that the confidence interval is

$$\left(\frac{(18-1) \times 6.48^2}{30.19}, \frac{(18-1) \times 6.48^2}{7.564} \right) = (23.6, 94.4).$$

At 99% confidence $\chi^2_{0.005,17} = 35.72$ and $\chi^2_{0.995,17} = 5.697$ so that the confidence interval is

$$\left(\frac{(18-1) \times 6.48^2}{35.72}, \frac{(18-1) \times 6.48^2}{5.697} \right) = (20.0, 125.3).$$

8.4.9   At 99% confidence $\chi^2_{0.005,40} = 66.77$ and $\chi^2_{0.995,40} = 20.71$ so that the confidence interval is

$$\left(\sqrt{\frac{(41-1) \times 0.124^2}{66.77}}, \sqrt{\frac{(41-1) \times 0.124^2}{20.71}} \right) = (0.095, 0.170).$$

8.4.10 At 95% confidence $\chi^2_{0.025,19} = 32.85$ and $\chi^2_{0.975,19} = 8.907$ so that the confidence interval is

$$\left(\frac{(20-1) \times 11.90^2}{32.85}, \frac{(20-1) \times 11.90^2}{8.907} \right) = (81.9, 302.1).$$

8.4.11 At 90% confidence $\chi^2_{0.05,15} = 25.00$ and $\chi^2_{0.95,15} = 7.261$ so that the confidence interval is

$$\left(\sqrt{\frac{(16-1) \times 0.058^2}{25.00}}, \sqrt{\frac{(16-1) \times 0.058^2}{7.261}} \right) = (0.045, 0.083).$$

At 95% confidence $\chi^2_{0.025,15} = 27.49$ and $\chi^2_{0.975,15} = 6.262$ so that the confidence interval is

$$\left(\sqrt{\frac{(16-1) \times 0.058^2}{27.49}}, \sqrt{\frac{(16-1) \times 0.058^2}{6.262}} \right) = (0.043, 0.090).$$

At 99% confidence $\chi^2_{0.005,15} = 32.80$ and $\chi^2_{0.995,15} = 4.601$ so that the confidence interval is

$$\left(\sqrt{\frac{(16-1) \times 0.058^2}{32.80}}, \sqrt{\frac{(16-1) \times 0.058^2}{4.601}} \right) = (0.039, 0.105).$$

Chapter 9

Comparing Two Population Means

9.2 Analysis of Paired Samples

9.2.1   The differences $z_i = x_i - y_i$ have a sample mean $\bar{z} = 7.12$ and a sample standard deviation $s = 34.12$.

Consider the hypothesis testing problem

$$H_0 : \mu = \mu_A - \mu_B \leq 0 \quad \text{versus} \quad H_A : \mu = \mu_A - \mu_B > 0$$

where the alternative hypothesis states that the new assembly method is quicker on average than the standard assembly method.

The test statistic is

$$t = \frac{\sqrt{n}\,\bar{z}}{s} = \frac{\sqrt{35} \times 7.12}{34.12} = 1.23.$$

The p-value is

$$P(X \geq 1.23) = 0.114$$

where the random variable X has a t distribution with $35 - 1 = 34$ degrees of freedom.

There is *not* sufficient evidence to conclude that the new assembly method is any quicker on average than the standard assembly method.

With $t_{0.05,34} = 1.691$ a one-sided 95% confidence level confidence interval for $\mu = \mu_A - \mu_B$ is

$$\left(7.12 - \frac{1.691 \times 34.12}{\sqrt{35}}, \infty\right) = (-2.63, \infty).$$

9.2.2   The differences $z_i = x_i - y_i$ have a sample mean $\bar{z} = -1.36$ and a sample standard deviation $s = 6.08$.

Consider the hypothesis testing problem

$$H_0 : \mu = \mu_A - \mu_B = 0 \quad \text{versus} \quad H_A : \mu = \mu_A - \mu_B \neq 0.$$

The test statistic is

$$t = \frac{\sqrt{n}\,\bar{z}}{s} = \frac{\sqrt{14} \times (-1.36)}{6.08} = -0.837.$$

The p-value is

$$2 \times P(X \leq -0.837) = 0.418$$

where the random variable X has a t distribution with $14-1 = 13$ degrees of freedom.

There is *not* sufficient evidence to conclude that the different stimulation conditions affect the adhesion of the red blood cells.

With $t_{0.025,13} = 2.160$ a two-sided 95% confidence level confidence interval for $\mu = \mu_A - \mu_B$ is

$$(-1.36 - \tfrac{2.160 \times 6.08}{\sqrt{14}}, -1.36 + \tfrac{2.160 \times 6.08}{\sqrt{14}}) = (-4.87, 2.15).$$

9.2.3   The differences $z_i = x_i - y_i$ have a sample mean $\bar{z} = 0.570$ and a sample standard deviation $s = 0.813$.

Consider the hypothesis testing problem

$$H_0 : \mu = \mu_A - \mu_B \leq 0 \quad \text{versus} \quad H_A : \mu = \mu_A - \mu_B > 0$$

where the alternative hypothesis states that the new tires have a smaller average reduction in tread depth than the standard tires.

The test statistic is

$$t = \frac{\sqrt{n}\,\bar{z}}{s} = \frac{\sqrt{20} \times 0.570}{0.813} = 3.14.$$

The p-value is

$$P(X \geq 3.14) = 0.003$$

where the random variable X has a t distribution with $20-1 = 19$ degrees of freedom.

There is sufficient evidence to conclude that the new tires are better than the standard tires in terms of the average reduction in tread depth.

With $t_{0.05,19} = 1.729$ a one-sided 95% confidence level confidence interval for $\mu = \mu_A - \mu_B$ is

$$(0.570 - \tfrac{1.729 \times 0.813}{\sqrt{20}}, \infty) = (0.256, \infty).$$

9.2.4   The differences $z_i = x_i - y_i$ have a sample mean $\bar{z} = -7.70$ and a sample standard deviation $s = 14.64$.

Consider the hypothesis testing problem

$$H_0 : \mu = \mu_A - \mu_B \geq 0 \quad \text{versus} \quad H_A : \mu = \mu_A - \mu_B < 0$$

where the alternative hypothesis states that the new teaching method produces higher scores on average than the standard teaching method.

The test statistic is

$$t = \frac{\sqrt{n}\,\bar{z}}{s} = \frac{\sqrt{40} \times (-7.70)}{14.64} = -3.33.$$

The p-value is

$$P(X \leq -3.33) = 0.001$$

where the random variable X has a t distribution with $40 - 1 = 39$ degrees of freedom.

There is sufficient evidence to conclude that the new teaching method is better since it produces higher scores on average than the standard teaching method.

With $t_{0.05,39} = 1.685$ a one-sided 95% confidence level confidence interval for $\mu = \mu_A - \mu_B$ is

$$\left(-\infty, -7.70 + \tfrac{1.685 \times 14.64}{\sqrt{40}}\right) = (-\infty, -3.80).$$

9.2.5   The differences $z_i = x_i - y_i$ have a sample mean $\bar{z} = 2.20$ and a sample standard deviation $s = 147.8$.

Consider the hypothesis testing problem

$$H_0 : \mu = \mu_A - \mu_B = 0 \quad \text{versus} \quad H_A : \mu = \mu_A - \mu_B \neq 0.$$

The test statistic is

$$t = \frac{\sqrt{n}\,\bar{z}}{s} = \frac{\sqrt{18} \times 2.20}{147.8} = 0.063.$$

The p-value is

$$2 \times P(X \geq 0.063) = 0.95$$

where the random variable X has a t distribution with $18 - 1 = 17$ degrees of freedom.

There is *not* sufficient evidence to conclude that the two laboratories are any different in the datings which they provide.

With $t_{0.025,17} = 2.110$ a two-sided 95% confidence level confidence interval for $\mu = \mu_A - \mu_B$ is

$$\left(2.20 - \tfrac{2.110 \times 147.8}{\sqrt{18}}, 2.20 + \tfrac{2.110 \times 147.8}{\sqrt{18}}\right) = (-71.3, 75.7).$$

9.2.6   The differences $z_i = x_i - y_i$ have a sample mean $\bar{z} = -1.42$ and a sample standard deviation $s = 12.74$.

Consider the hypothesis testing problem

$$H_0 : \mu = \mu_A - \mu_B \geq 0 \quad \text{versus} \quad H_A : \mu = \mu_A - \mu_B < 0$$

where the alternative hypothesis states that the new golf balls travel further on average than the standard golf balls.

The test statistic is

$$t = \frac{\sqrt{n}\,\bar{z}}{s} = \frac{\sqrt{24} \times (-1.42)}{12.74} = -0.546.$$

The p-value is

$$P(X \leq -0.546) = 0.30$$

where the random variable X has a t distribution with $24-1 = 23$ degrees of freedom.

There is *not* sufficient evidence to conclude that the new golf balls travel further on average than the standard golf balls.

With $t_{0.05,23} = 1.714$ a one-sided 95% confidence level confidence interval for $\mu = \mu_A - \mu_B$ is

$$(-\infty, -1.42 + \tfrac{1.714 \times 12.74}{\sqrt{24}}) = (-\infty, 3.04).$$

9.3 Analysis of Independent Samples

9.3.2   (a) The pooled variance is

$$s_p^2 = \frac{(n-1)s_x^2 + (m-1)s_y^2}{n+m-2} = \frac{((14-1) \times 4.30^2) + ((14-1) \times 5.23^2)}{14+14-2}$$

$$= 22.92.$$

With $t_{0.005,26} = 2.779$ a 99% two-sided confidence interval for $\mu_A - \mu_B$ is

$$(32.45 - 41.45 - 2.779 \times \sqrt{22.92} \times \sqrt{\tfrac{1}{14} + \tfrac{1}{14}},$$

$$32.45 - 41.45 + 2.779 \times \sqrt{22.92} \times \sqrt{\tfrac{1}{14} + \tfrac{1}{14}}) = (-14.03, -3.97).$$

(b) With degrees of freedom

$$\nu = \frac{\left(\frac{4.30^2}{14} + \frac{5.23^2}{14}\right)^2}{\frac{4.30^4}{14^2 \times (14-1)} + \frac{5.23^4}{14^2 \times (14-1)}} = 25.06$$

and consequently a critical point $t_{0.005,25} = 2.787$, a 99% two-sided confidence interval for $\mu_A - \mu_B$ is

$$(32.45 - 41.45 - 2.787 \times \sqrt{\tfrac{4.30^2}{14} + \tfrac{5.23^2}{14}}, 32.45 - 41.45 + 2.787 \times \sqrt{\tfrac{4.30^2}{14} + \tfrac{5.23^2}{14}})$$

$$= (-14.04, -3.96).$$

(c) The test statistic is

$$t = \frac{\bar{x} - \bar{y}}{\sqrt{\frac{s_x^2}{n} + \frac{s_y^2}{m}}} = \frac{32.45 - 41.45}{\sqrt{\frac{4.30^2}{14} + \frac{5.23^2}{14}}} = 4.97.$$

The null hypothesis is rejected since $|t| = 4.97$ is larger than the critical point $t_{0.005,26} = 2.779$.

The p-value is

$$2 \times P(X \geq 4.97) = 0.000$$

where the random variable X has a t distribution with $14 + 14 - 2 = 26$ degrees of freedom.

9.3.3   (a) The pooled variance is

$$s_p^2 = \frac{(n-1)s_x^2 + (m-1)s_y^2}{n+m-2} = \frac{((8-1) \times 44.76^2) + ((17-1) \times 38.94^2)}{8+17-2}$$

$$= 1,664.6.$$

With $t_{0.005,23} = 2.807$ a 99% two-sided confidence interval for $\mu_A - \mu_B$ is

$$(675.1 - 702.4 - 2.807 \times \sqrt{1664.6} \times \sqrt{\tfrac{1}{8} + \tfrac{1}{17}},$$

$$675.1 - 702.4 + 2.807 \times \sqrt{1664.6} \times \sqrt{\tfrac{1}{8} + \tfrac{1}{17}}) = (-76.4, 21.8).$$

(b) With degrees of freedom

$$\nu = \frac{\left(\frac{44.76^2}{8} + \frac{38.94^2}{17}\right)^2}{\frac{44.76^4}{8^2 \times (8-1)} + \frac{38.94^4}{17^2 \times (17-1)}} = 12.2$$

and consequently a critical point $t_{0.005,12} = 3.055$, a 99% two-sided confidence interval for $\mu_A - \mu_B$ is

$$(675.1 - 702.4 - 3.055 \times \sqrt{\frac{44.76^2}{8} + \frac{38.94^2}{17}}, 675.1 - 702.4 + 3.055 \times \sqrt{\frac{44.76^2}{8} + \frac{38.94^2}{17}})$$
$$= (-83.6, 29.0).$$

(c) The test statistic is

$$t = \frac{\bar{x} - \bar{y}}{s_p\sqrt{\frac{1}{n} + \frac{1}{m}}} = \frac{675.1 - 702.4}{\sqrt{1664.6} \times \sqrt{\frac{1}{8} + \frac{1}{17}}} = -1.56.$$

The null hypothesis is accepted since $|t| = 1.56$ is smaller than the critical point $t_{0.005,23} = 2.807$.

The p-value is

$$2 \times P(X \geq 1.56) = 0.132$$

where the random variable X has a t distribution with $8 + 17 - 2 = 23$ degrees of freedom.

9.3.4 (a) With degrees of freedom

$$\nu = \frac{\left(\frac{1.07^2}{10} + \frac{0.62^2}{9}\right)^2}{\frac{1.07^4}{10^2 \times (10-1)} + \frac{0.62^4}{9^2 \times (9-1)}} = 14.7$$

and consequently a critical point $t_{0.01,14} = 2.624$, a 99% one-sided confidence interval for $\mu_A - \mu_B$ is

$$(7.76 - 6.88 - 2.624 \times \sqrt{\frac{1.07^2}{10} + \frac{0.62^2}{9}}, \infty) = (-0.16, \infty).$$

(b) The value of c increases with a confidence level of 95%.

(c) The test statistic is

$$t = \frac{\bar{x} - \bar{y}}{\sqrt{\frac{s_x^2}{n} + \frac{s_y^2}{m}}} = \frac{7.76 - 6.88}{\sqrt{\frac{1.07^2}{10} + \frac{0.62^2}{9}}} = 2.22.$$

The null hypothesis is accepted since $t = 2.22 \leq t_{0.01,14} = 2.624$.

The p-value is

$$P(X \geq 2.22) = 0.022$$

where the random variable X has a t distribution with 14 degrees of freedom.

9.3.5   (a)   The pooled variance is

$$s_p^2 = \frac{(n-1)s_x^2 + (m-1)s_y^2}{n+m-2}$$

$$= \frac{((13-1) \times 0.00128^2) + ((15-1) \times 0.00096^2)}{13+15-2} = 1.25 \times 10^{-6}.$$

With $t_{0.05,26} = 1.706$ a 95% one-sided confidence interval for $\mu_A - \mu_B$ is

$(-\infty, 0.0548 - 0.0569 + 1.706 \times \sqrt{1.25 \times 10^{-6}} \times \sqrt{\frac{1}{13} + \frac{1}{15}})$

$= (-\infty, -0.0014).$

(b)   The test statistic is

$$t = \frac{\bar{x} - \bar{y}}{s_p\sqrt{\frac{1}{n} + \frac{1}{m}}} = \frac{0.0548 - 0.0569}{\sqrt{1.25 \times 10^{-6}} \times \sqrt{\frac{1}{13} + \frac{1}{15}}} = -4.95.$$

The null hypothesis is rejected at size $\alpha = 0.01$ since $t = -4.95 < -t_{0.01,26} = -2.479$. The null hypothesis is consequently also rejected at size $\alpha = 0.05$.
The p-value is

$$P(X \le -4.95) = 0.000$$

where the random variable X has a t distribution with $13 + 15 - 2 = 26$ degrees of freedom.

9.3.6   (a)   The pooled variance is

$$s_p^2 = \frac{(n-1)s_x^2 + (m-1)s_y^2}{n+m-2} = \frac{((41-1) \times 0.124^2) + ((41-1) \times 0.137^2)}{41+41-2}$$

$$= 0.01707.$$

The test statistic is

$$t = \frac{\bar{x} - \bar{y}}{s_p\sqrt{\frac{1}{n} + \frac{1}{m}}} = \frac{3.04 - 3.12}{\sqrt{0.01707} \times \sqrt{\frac{1}{41} + \frac{1}{41}}} = -2.77.$$

The null hypothesis is rejected at size $\alpha = 0.01$ since $|t| = 2.77$ is larger than $t_{0.005,80} = 2.639$.
The p-value is

$$2 \times P(X \le -2.77) = 0.007$$

where the random variable X has a t distribution with $41 + 41 - 2 = 80$ degrees of freedom.

(b)   With $t_{0.005,80} = 2.639$ a 99% two-sided confidence interval for $\mu_A - \mu_B$ is

$(3.04 - 3.12 - 2.639 \times \sqrt{0.01707} \times \sqrt{\frac{1}{41} + \frac{1}{41}},$

$3.04 - 3.12 + 2.639 \times \sqrt{0.01707} \times \sqrt{\frac{1}{41} + \frac{1}{41}}) = (-0.156, -0.004).$

(c) There is sufficient evidence to conclude that the average thicknesses of sheets produced by the two processes are different.

9.3.7   (a) The appropriate degrees of freedom are

$$\nu = \frac{\left(\frac{11.90^2}{20} + \frac{4.61^2}{25}\right)^2}{\frac{11.90^4}{20^2 \times (20-1)} + \frac{4.61^4}{25^2 \times (25-1)}} = 23.6.$$

Consider the hypothesis testing problem

$$H_0 : \mu = \mu_A - \mu_B \geq 0 \quad \text{versus} \quad H_A : \mu = \mu_A - \mu_B < 0$$

where the alternative hypothesis states that the synthetic fiber bundles have an average breaking strength larger than the wool fiber bundles.

The test statistic is

$$t = \frac{\bar{x} - \bar{y}}{\sqrt{\frac{s_x^2}{n} + \frac{s_y^2}{m}}} = \frac{436.5 - 452.8}{\sqrt{\frac{11.90^2}{20} + \frac{4.61^2}{25}}} = -5.788.$$

The null hypothesis is rejected at size $\alpha = 0.01$ since $t = -5.788 < -t_{0.01,23} = -2.500$.

The p-value is

$$P(X \leq -5.788) = 0.000$$

where the random variable X has a t distribution with 23 degrees of freedom.

(b) With a critical point $t_{0.01,23} = 2.500$, a 99% one-sided confidence interval for $\mu_A - \mu_B$ is

$$(-\infty, 436.5 - 452.8 + 2.500 \times \sqrt{\tfrac{11.90^2}{20} + \tfrac{4.61^2}{25}}) = (-\infty, -9.3).$$

(c) There is sufficient evidence to conclude that the synthetic fiber bundles have an average breaking strength larger than the wool fiber bundles.

9.3.8   For a general analysis without assuming equal population variances the appropriate degrees of freedom are

$$\nu = \frac{\left(\frac{0.058^2}{16} + \frac{0.062^2}{16}\right)^2}{\frac{0.058^4}{16^2 \times (16-1)} + \frac{0.062^4}{16^2 \times (16-1)}} = 29.9.$$

Consider the hypothesis testing problem

$$H_0 : \mu = \mu_A - \mu_B \geq 0 \quad \text{versus} \quad H_A : \mu = \mu_A - \mu_B < 0$$

where the alternative hypothesis states that the brand B sugar packets weigh slightly more on average than brand A sugar packets.

The test statistic is

$$t = \frac{\bar{x} - \bar{y}}{\sqrt{\frac{s_x^2}{n} + \frac{s_y^2}{m}}} = \frac{1.053 - 1.071}{\sqrt{\frac{0.058^2}{16} + \frac{0.062^2}{16}}} = -0.848.$$

The p-value is

$$P(X \leq -0.848) = 0.798$$

where the random variable X has a t distribution with 29 degrees of freedom.

There is *not* sufficient evidence to conclude that the brand B sugar packets weigh slightly more on average than brand A sugar packets.

9.3.9   (a) The test statistic is

$$z = \frac{\bar{x} - \bar{y} - \delta}{\sqrt{\frac{\sigma_A^2}{n} + \frac{\sigma_B^2}{m}}} = \frac{100.85 - 89.32 - 3}{\sqrt{\frac{25^2}{47} + \frac{20^2}{62}}} = 1.92.$$

The p-value is

$$2 \times \Phi(-1.92) = 0.055.$$

(b) With a critical point $z_{0.05} = 1.645$ a 90% two-sided confidence interval for $\mu_A - \mu_B$ is

$$(100.85 - 89.32 - 1.645 \times \sqrt{\tfrac{25^2}{47} + \tfrac{20^2}{62}}, 100.85 - 89.32 + 1.645 \times \sqrt{\tfrac{25^2}{47} + \tfrac{20^2}{62}})$$
$$= (4.22, 18.84).$$

9.3.10   (a) The test statistic is

$$z = \frac{\bar{x} - \bar{y}}{\sqrt{\frac{\sigma_A^2}{n} + \frac{\sigma_B^2}{m}}} = \frac{5.782 - 6.443}{\sqrt{\frac{2.0^2}{38} + \frac{2.0^2}{40}}} = -1.459.$$

The p-value is

$$\Phi(-1.459) = 0.072.$$

(b) With a critical point $z_{0.01} = 2.326$ a 99% one-sided confidence interval for $\mu_A - \mu_B$ is

$$(-\infty, 5.782 - 6.443 + 2.326 \times \sqrt{\tfrac{2.0^2}{38} + \tfrac{2.0^2}{40}}) = (-\infty, 0.393).$$

9.3.11   (a) The test statistic is

$$z = \frac{\bar{x} - \bar{y}}{\sqrt{\frac{\sigma_A^2}{n} + \frac{\sigma_B^2}{m}}} = \frac{19.50 - 18.64}{\sqrt{\frac{1.0^2}{10} + \frac{1.0^2}{12}}} = 2.009.$$

The p-value is

$$2 \times \Phi(-2.009) = 0.045.$$

(b) With a critical point $z_{0.05} = 1.645$ a 90% two-sided confidence interval for $\mu_A - \mu_B$ is

$$(19.50 - 18.64 - 1.645 \times \sqrt{\tfrac{1.0^2}{10} + \tfrac{1.0^2}{12}}, 19.50 - 18.64 + 1.645 \times \sqrt{\tfrac{1.0^2}{10} + \tfrac{1.0^2}{12}})$$
$$= (0.16, 1.56).$$

With a critical point $z_{0.025} = 1.960$ a 95% two-sided confidence interval for $\mu_A - \mu_B$ is

$$(19.50 - 18.64 - 1.960 \times \sqrt{\tfrac{1.0^2}{10} + \tfrac{1.0^2}{12}}, 19.50 - 18.64 + 1.960 \times \sqrt{\tfrac{1.0^2}{10} + \tfrac{1.0^2}{12}})$$
$$= (0.02, 1.70).$$

With a critical point $z_{0.005} = 2.576$ a 99% two-sided confidence interval for $\mu_A - \mu_B$ is

$$(19.50 - 18.64 - 2.576 \times \sqrt{\tfrac{1.0^2}{10} + \tfrac{1.0^2}{12}}, 19.50 - 18.64 + 2.576 \times \sqrt{\tfrac{1.0^2}{10} + \tfrac{1.0^2}{12}})$$
$$= (-0.24, 1.96).$$

9.3.12 Using 2.6 as an upper bound for $t_{0.005,\nu}$, equal sample sizes of

$$n = m \geq \frac{4\, t_{\alpha/2,\nu}^2\, (\sigma_A^2 + \sigma_B^2)}{L_0^2} = \frac{4 \times 2.6^2 \times (10.0^2 + 15.0^2)}{10.0^2} = 87.88$$

should be sufficient. Equal sample sizes of at least 88 can be recommended.

9.3.13 Using 2.0 as an upper bound for $t_{0.025,\nu}$, equal sample sizes of

$$n = m \geq \frac{4\, t_{\alpha/2,\nu}^2\, (\sigma_A^2 + \sigma_B^2)}{L_0^2} = \frac{4 \times 2.0^2 \times (1.2^2 + 1.2^2)}{1.0^2} = 46.08$$

should be sufficient. Equal sample sizes of at least 47 can be recommended.

9.3.14 Using $t_{0.005,26} = 2.779$ equal total sample sizes of

$$n = m \geq \frac{4\, t_{\alpha/2,\nu}^2\, (s_x^2 + s_y^2)}{L_0^2} = \frac{4 \times 2.779^2 \times (4.30^2 + 5.23^2)}{5.0^2} = 56.6$$

should be sufficient. Additional sample sizes of at least $57 - 14 = 43$ from each population can be recommended.

9.3.15   Using $t_{0.005,80} = 2.639$ equal total sample sizes of

$$n = m \geq \frac{4\, t_{\alpha/2,\nu}^2\, (s_x^2 + s_y^2)}{L_0^2} = \frac{4 \times 2.639^2 \times (0.124^2 + 0.137^2)}{0.1^2} = 95.1$$

should be sufficient. Additional sample sizes of at least $96 - 41 = 55$ from each population can be recommended.

9.3.16   There is *not* sufficient evidence to conclude that the average service times are any different at these two times of day.

9.3.17   There is sufficient evidence to conclude that the bricks from company A weigh more on average than the bricks from company B. There is also more variability in the weights of the bricks from company A.

9.3.18   There is a fairly strong suggestion that the paint thicknesses from production line A are larger than those from production line B although the evidence is not completely overwhelming (p-value of 0.011).

9.3.19   There is sufficient evidence to conclude that the damped feature is effective in reducing heel-strike force.

9.5 Supplementary Problems

9.5.1 The differences $z_i = x_i - y_i$ have a sample mean $\bar{z} = 2.85$ and a sample standard deviation $s = 5.30$.

Consider the hypothesis testing problem

$$H_0 : \mu = \mu_A - \mu_B \leq 0 \quad \text{versus} \quad H_A : \mu = \mu_A - \mu_B > 0$$

where the alternative hypothesis states that the color displays are more effective than the black and white displays.

The test statistic is

$$t = \frac{\sqrt{n}\,\bar{z}}{s} = \frac{\sqrt{22} \times 2.85}{5.30} = 2.52.$$

The p-value is

$$P(X \geq 2.52) = 0.010$$

where the random variable X has a t distribution with $22-1 = 21$ degrees of freedom.

There is sufficient evidence to conclude that the color displays are more effective than the black and white displays.

With $t_{0.05,21} = 1.721$ a one-sided 95% confidence level confidence interval for $\mu = \mu_A - \mu_B$ is

$$(2.85 - \tfrac{1.721 \times 5.30}{\sqrt{22}}, \infty) = (0.91, \infty).$$

9.5.2 The differences $z_i = x_i - y_i$ have a sample mean $\bar{z} = 7.50$ and a sample standard deviation $s = 6.84$.

Consider the hypothesis testing problem

$$H_0 : \mu = \mu_A - \mu_B = 0 \quad \text{versus} \quad H_A : \mu = \mu_A - \mu_B \neq 0.$$

The test statistic is

$$t = \frac{\sqrt{n}\,\bar{z}}{s} = \frac{\sqrt{14} \times 7.50}{6.84} = 4.10.$$

The p-value is

$$2 \times P(X \geq 4.10) = 0.001$$

where the random variable X has a t distribution with $14-1 = 13$ degrees of freedom.

There is sufficient evidence to conclude that the water absorption properties of the fabric are different for the two different roller pressures.

With $t_{0.025,13} = 2.160$ a two-sided 95% confidence level confidence interval for $\mu = \mu_A - \mu_B$ is

$$(7.50 - \tfrac{2.160 \times 6.84}{\sqrt{14}}, 7.50 + \tfrac{2.160 \times 6.84}{\sqrt{14}}) = (3.55, 11.45).$$

9.5.3 (a) The appropriate degrees of freedom are

$$\nu = \frac{\left(\frac{5.20^2}{35} + \frac{3.06^2}{35}\right)^2}{\frac{5.20^4}{35^2 \times (35-1)} + \frac{3.06^4}{35^2 \times (35-1)}} = 55.03.$$

The test statistic is

$$t = \frac{\bar{x} - \bar{y}}{\sqrt{\frac{s_x^2}{n} + \frac{s_y^2}{m}}} = \frac{22.73 - 12.66}{\sqrt{\frac{5.20^2}{35} + \frac{3.06^2}{35}}} = 9.87.$$

The p-value is

$$2 \times P(X \geq 9.87) = 0.000$$

where the random variable X has a t distribution with 55 degrees of freedom. It is not plausible that the average crystal size does not depend upon the pre-expansion temperature.

(b) With a critical point $t_{0.005,55} = 2.668$ a 99% two-sided confidence interval for $\mu_A - \mu_B$ is

$$(22.73 - 12.66 - 2.668 \times \sqrt{\frac{5.20^2}{35} + \frac{3.06^2}{35}}, 22.73 - 12.66 + 2.668 \times \sqrt{\frac{5.20^2}{35} + \frac{3.06^2}{35}})$$
$$= (7.35, 12.79).$$

(c) Using $t_{0.005,55} = 2.668$ equal total sample sizes of

$$n = m \geq \frac{4\, t_{\alpha/2,\nu}^2\, (s_x^2 + s_y^2)}{L_0^2} = \frac{4 \times 2.668^2 \times (5.20^2 + 3.06^2)}{4.0^2} = 64.8$$

should be sufficient. Additional sample sizes of at least $65 - 35 = 30$ from each population can be recommended.

9.5.4 For a general analysis without assuming equal population variances the appropriate degrees of freedom are

$$\nu = \frac{\left(\frac{20.39^2}{48} + \frac{15.62^2}{10}\right)^2}{\frac{20.39^4}{48^2 \times (48-1)} + \frac{15.62^4}{10^2 \times (10-1)}} = 16.1.$$

Consider the hypothesis testing problem

$$H_0 : \mu = \mu_A - \mu_B \leq 0 \quad \text{versus} \quad H_A : \mu = \mu_A - \mu_B > 0$$

where the alternative hypothesis states that the new driving route is quicker on average than the standard driving route.

The test statistic is

$$t = \frac{\bar{x} - \bar{y}}{\sqrt{\frac{s_x^2}{n} + \frac{s_y^2}{m}}} = \frac{432.7 - 403.5}{\sqrt{\frac{20.39^2}{48} + \frac{15.62^2}{10}}} = 5.08.$$

The p-value is

$$P(X \geq 5.08) = 0.000$$

where the random variable X has a t distribution with 16 degrees of freedom.

There is sufficient evidence to conclude that the new driving route is quicker on average than the standard driving route.

9.5.5   There is sufficient evidence to conclude that the additional sunlight results in larger heights on average.

9.5.6   There is *not* sufficient evidence to conclude that the reorganization has produced any improvement in the average waiting time. However, the variability in the waiting times has been reduced following the reorganization.

9.5.8   With $F_{0.05,17,20} = 2.1667$ and $F_{0.05,20,17} = 2.2304$ the confidence interval is

$$\left(\frac{6.48^2}{9.62^2 \times 2.1667}, \frac{6.48^2 \times 2.2304}{9.62^2} \right) = (0.21, 1.01).$$

9.5.9   With $F_{0.05,40,40} = 1.6928$ the confidence interval is

$$\left(\frac{0.124^2}{0.137^2 \times 1.6928}, \frac{0.124^2 \times 1.6928}{0.137^2} \right) = (0.484, 1.387).$$

9.5.10   With $F_{0.05,19,24} = 2.0399$ and $F_{0.05,24,19} = 2.1141$ the 90% confidence interval is

$$\left(\frac{11.90^2}{4.61^2 \times 2.0399}, \frac{11.90^2 \times 2.1141}{4.61^2} \right) = (3.27, 14.09).$$

With $F_{0.025,19,24} = 2.3452$ and $F_{0.025,24,19} = 2.4523$ the 95% confidence interval is

$$\left(\frac{11.90^2}{4.61^2 \times 2.3452}, \frac{11.90^2 \times 2.4523}{4.61^2} \right) = (2.84, 16.34).$$

With $F_{0.005,19,24} = 3.0920$ and $F_{0.005,24,19} = 3.3062$ the 99% confidence interval is

$$\left(\frac{11.90^2}{4.61^2 \times 3.0920}, \frac{11.90^2 \times 3.3062}{4.61^2} \right) = (2.16, 22.03).$$

Chapter 10

Discrete Data Analysis

10.1 Inferences on a Population Proportion

10.1.1   (a) With $z_{0.005} = 2.576$ the confidence interval is

$$\left(\frac{11}{32} - \frac{2.576}{32} \times \sqrt{\frac{11 \times (32 - 11)}{32}}, \frac{11}{32} + \frac{2.576}{32} \times \sqrt{\frac{11 \times (32 - 11)}{32}} \right)$$

$$= (0.127, 0.560).$$

  (b) With $z_{0.025} = 1.960$ the confidence interval is

$$\left(\frac{11}{32} - \frac{1.960}{32} \times \sqrt{\frac{11 \times (32 - 11)}{32}}, \frac{11}{32} + \frac{1.960}{32} \times \sqrt{\frac{11 \times (32 - 11)}{32}} \right)$$

$$= (0.179, 0.508).$$

  (c) With $z_{0.01} = 2.326$ the confidence interval is

$$\left(0, \frac{11}{32} + \frac{2.326}{32} \times \sqrt{\frac{11 \times (32 - 11)}{32}} \right) = (0, 0.539).$$

  (d) The exact p-value is

$$2 \times P(X \leq 11) = 0.110$$

where the random variable X has a $B(32, 0.5)$ distribution.
The statistic for the normal approximation of the p-value is

$$z = \frac{x - np_0}{\sqrt{np_0(1 - p_0)}} = \frac{11 - (32 \times 0.5)}{\sqrt{32 \times 0.5 \times (1 - 0.5)}} = -1.768$$

and the p-value is

$$2 \times \Phi(-1.768) = 0.077.$$

10.1.2   (a) With $z_{0.005} = 2.576$ the confidence interval is

$$\left(\frac{21}{27} - \frac{2.576}{27} \times \sqrt{\frac{21 \times (27 - 21)}{27}}, \frac{21}{27} + \frac{2.576}{27} \times \sqrt{\frac{21 \times (27 - 21)}{27}} \right)$$

$$= (0.572, 0.984).$$

(b) With $z_{0.025} = 1.960$ the confidence interval is

$$\left(\frac{21}{27} - \frac{1.960}{27} \times \sqrt{\frac{21 \times (27 - 21)}{27}}, \frac{21}{27} + \frac{1.960}{27} \times \sqrt{\frac{21 \times (27 - 21)}{27}} \right)$$

$$= (0.621, 0.935).$$

(c) With $z_{0.05} = 1.645$ the confidence interval is

$$\left(\frac{21}{27} - \frac{1.645}{27} \times \sqrt{\frac{21 \times (27 - 21)}{27}}, 1 \right) = (0.646, 1).$$

(d) The exact p-value is

$$P(X \geq 21) = 0.042$$

where the random variable X has a $B(27, 0.6)$ distribution.
The statistic for the normal approximation of the p-value is

$$z = \frac{x - np_0}{\sqrt{np_0(1 - p_0)}} = \frac{21 - (27 \times 0.6)}{\sqrt{27 \times 0.6 \times (1 - 0.6)}} = 1.886$$

and the p-value is

$$1 - \Phi(1.886) = 0.030.$$

10.1.3   (a) Let p be the probability that a value produced by the random number generator is a zero and consider the hypothesis testing problem

$$H_0 : p = 0.5 \quad \text{versus} \quad H_A : p \neq 0.5$$

where the alternative hypothesis states that the random number generator is producing 0's and 1's with unequal probabilities.
The statistic for the normal approximation of the p-value is

$$z = \frac{x - np_0}{\sqrt{np_0(1 - p_0)}} = \frac{25,264 - (50,000 \times 0.5)}{\sqrt{50,000 \times 0.5 \times (1 - 0.5)}} = 2.361$$

and the p-value is

$$2 \times \Phi(-2.361) = 0.018.$$

There is a fairly strong suggestion that the random number generator is producing 0's and 1's with unequal probabilities although the evidence is not completely overwhelming.

(b) With $z_{0.005} = 2.576$ the confidence interval is

$$\left(\frac{25{,}264}{50{,}000} - \frac{2.576}{50{,}000} \times \sqrt{\frac{25{,}264 \times (50{,}000 - 25{,}264)}{50{,}000}} , \right.$$

$$\left. \frac{25{,}264}{50{,}000} + \frac{2.576}{50{,}000} \times \sqrt{\frac{25{,}264 \times (50{,}000 - 25{,}264)}{50{,}000}} \right)$$

$$= (0.4995, 0.5110).$$

(c) Using the worst case scenario $\hat{p}(1 - \hat{p}) = 0.25$ the total sample size required can be calculated as

$$n \geq \frac{4 \, z_{\alpha/2}^2 \, \hat{p}(1 - \hat{p})}{L^2} = \frac{4 \times 2.576^2 \times 0.25}{0.005^2} = 265{,}431.04$$

so that an additional sample size of $265{,}432 - 50{,}000 \simeq 215{,}500$ would be required.

10.1.4 With $z_{0.05} = 1.645$ the confidence interval is

$$\left(\frac{35}{44} - \frac{1.645}{44} \times \sqrt{\frac{35 \times (44 - 35)}{44}} , 1 \right) = (0.695, 1).$$

10.1.5 Let p be the probability that a six is scored on the die and consider the hypothesis testing problem

$$H_0 : p \geq \frac{1}{6} \quad \text{versus} \quad H_A : p < \frac{1}{6}$$

where the alternative hypothesis states that the die has been weighted to reduce the chance of scoring a six.

In the first experiment the exact p-value is

$$P(X \leq 2) = 0.0066$$

where the random variable X has a $B(50, 1/6)$ distribution, and in the second experiment the exact p-value is

$$P(X \leq 4) = 0.0001$$

where the random variable X has a $B(100, 1/6)$ distribution. There is more support for foul play from the second experiment than from the first.

10.1.6   The exact p-value is

$$2 \times P(X \geq 21) = 0.304$$

where the random variable X has a $B(100, 1/6)$ distribution. The null hypothesis is accepted at the size $\alpha = 0.05$ level.

10.1.7   Let p be the probability that a juror is selected from the county where the investigator lives and consider the hypothesis testing problem

$$H_0 : p = 0.14 \quad \text{versus} \quad H_A : p \neq 0.14$$

where the alternative hypothesis implies that the jurors are not being randomly selected.

The statistic for the normal approximation of the p-value is

$$z = \frac{x - np_0}{\sqrt{np_0(1 - p_0)}} = \frac{122 - (1,386 \times 0.14)}{\sqrt{1,386 \times 0.14 \times (1 - 0.14)}} = -5.577$$

and the p-value is

$$2 \times \Phi(-5.577) = 0.000.$$

There is sufficient evidence to conclude that the jurors are not being randomly selected.

10.1.8   The statistic for the normal approximation of the p-value is

$$z = \frac{x - np_0}{\sqrt{np_0(1 - p_0)}} = \frac{23 - (324 \times 0.1)}{\sqrt{324 \times 0.1 \times (1 - 0.1)}} = -1.741$$

and the p-value is

$$\Phi(-1.741) = 0.041.$$

With $z_{0.01} = 2.326$ the confidence interval is

$$\left(0, \frac{23}{324} + \frac{2.326}{324} \times \sqrt{\frac{23 \times (324 - 23)}{324}} \right) = (0, 0.104).$$

It has not been conclusively shown that the screeing test is acceptable.

10.1.9   With $z_{0.025} = 1.960$ and $L = 0.02$, for the worst case scenario with $\hat{p}(1 - \hat{p}) = 0.25$ the required sample size can be calculated as

$$n \geq \frac{4\,z_{\alpha/2}^2\,\hat{p}(1 - \hat{p})}{L^2} = \frac{4 \times 1.960^2 \times 0.25}{0.02^2} = 9{,}604.$$

If it can be assumed that $\hat{p}(1 - \hat{p}) \leq 0.75 \times 0.25 = 0.1875$ then the required sample size can be calculated as

$$n \geq \frac{4\,z_{\alpha/2}^2\,\hat{p}(1 - \hat{p})}{L^2} = \frac{4 \times 1.960^2 \times 0.1875}{0.02^2} = 7{,}203.$$

10.1.10   With $z_{0.005} = 2.576$ and $L = 0.04$, for the worst case scenario with $\hat{p}(1 - \hat{p}) = 0.25$ the required sample size can be calculated as

$$n \geq \frac{4\,z_{\alpha/2}^2\,\hat{p}(1 - \hat{p})}{L^2} = \frac{4 \times 2.576^2 \times 0.25}{0.04^2} = 4{,}148.$$

If it can be assumed that $\hat{p}(1 - \hat{p}) \leq 0.4 \times 0.6 = 0.24$ then the required sample size can be calculated as

$$n \geq \frac{4\,z_{\alpha/2}^2\,\hat{p}(1 - \hat{p})}{L^2} = \frac{4 \times 2.576^2 \times 0.24}{0.04^2} = 3{,}982.$$

10.1.11   With $z_{0.005} = 2.576$ the confidence interval is

$$\left(\frac{73}{120} - \frac{2.576}{120} \times \sqrt{\frac{73 \times (120 - 73)}{120}}, \frac{73}{120} + \frac{2.576}{120} \times \sqrt{\frac{73 \times (120 - 73)}{120}} \right)$$

$$= (0.494, 0.723).$$

Using

$$\hat{p}(1 - \hat{p}) = \frac{73}{120} \times \left(1 - \frac{73}{120} \right) = 0.238$$

the total sample size required can be calculated as

$$n \geq \frac{4\,z_{\alpha/2}^2\,\hat{p}(1 - \hat{p})}{L^2} = \frac{4 \times 2.576^2 \times 0.238}{0.1^2} = 631.7$$

so that an additional sample size of $632 - 120 = 512$ would be required.

10.1.12   Let p be the proportion of defective chips in the shipment. With $z_{0.05} = 1.645$ a 95% upper confidence bound on p is

$$\left(0 \, , \, \frac{6}{200} + \frac{1.645}{200} \times \sqrt{\frac{6 \times (200 - 6)}{200}} \right) = (0, 0.049842).$$

A 95% upper confidence bound on the total number of defective chips in the shipment can therefore be calculated as $0.049842 \times 100,000 = 4,984.2$ or 4,985 chips.

10.1.13   With $z_{0.025} = 1.960$ the confidence interval is

$$\left(\frac{12}{20} - \frac{1.960}{20} \times \sqrt{\frac{12 \times (20 - 12)}{20}} , \frac{12}{20} + \frac{1.960}{20} \times \sqrt{\frac{12 \times (20 - 12)}{20}} \right)$$

$$= (0.385, 0.815).$$

10.1.14   Let p be the proportion of the applications which contained errors. With $z_{0.05} = 1.645$ a 95% lower confidence bound on p is

$$\left(\frac{17}{85} - \frac{1.645}{85} \times \sqrt{\frac{17 \times (85 - 17)}{85}} , 1 \right) = (0.1286, 1).$$

A 95% lower confidence bound on the total number of applications which contained errors can therefore be calculated as $0.1286 \times 7,607 = 978.5$ or 979 applications.

10.1.15   With $z_{0.025} = 1.960$ and $L = 0.10$, for the worst case scenario with $\hat{p}(1 - \hat{p}) = 0.25$ the required sample size can be calculated as

$$n \geq \frac{4 \, z_{\alpha/2}^2 \, \hat{p}(1 - \hat{p})}{L^2} = \frac{4 \times 1.960^2 \times 0.25}{0.10^2} = 384.2$$

or 385 householders.

If it can be assumed that $\hat{p}(1 - \hat{p}) \leq 0.333 \times 0.667 = 0.222$ then the required sample size can be calculated as

$$n \geq \frac{4 \, z_{\alpha/2}^2 \, \hat{p}(1 - \hat{p})}{L^2} = \frac{4 \times 1.960^2 \times 0.222}{0.10^2} = 341.1$$

or 342 householders.

10.1.16 With $z_{0.005} = 2.576$ the confidence interval is

$$\left(\frac{22}{542} - \frac{2.576}{542} \times \sqrt{\frac{22 \times (542 - 22)}{542}}, \frac{22}{542} + \frac{2.576}{542} \times \sqrt{\frac{22 \times (542 - 22)}{542}} \right)$$

$$= (0.019, 0.062).$$

10.1.17 The standard confidence interval is $(0.161, 0.557)$.

The alternative confidence interval is $(0.195, 0.564)$.

10.2   Comparing Two Population Proportions

10.2.1   (a) With $z_{0.005} = 2.576$ the confidence interval is

$$\left(\frac{14}{37} - \frac{7}{26} - 2.576 \times \sqrt{ \frac{14 \times (37 - 14)}{37^3} + \frac{7 \times (26 - 7)}{26^3} } \right. ,$$

$$\left. \frac{14}{37} - \frac{7}{26} + 2.576 \times \sqrt{ \frac{14 \times (37 - 14)}{37^3} + \frac{7 \times (26 - 7)}{26^3} } \right)$$

$$= (-0.195, 0.413).$$

(b) With $z_{0.025} = 1.960$ the confidence interval is

$$\left(\frac{14}{37} - \frac{7}{26} - 1.960 \times \sqrt{ \frac{14 \times (37 - 14)}{37^3} + \frac{7 \times (26 - 7)}{26^3} } \right. ,$$

$$\left. \frac{14}{37} - \frac{7}{26} + 1.960 \times \sqrt{ \frac{14 \times (37 - 14)}{37^3} + \frac{7 \times (26 - 7)}{26^3} } \right)$$

$$= (-0.122, 0.340).$$

(c) With $z_{0.01} = 2.326$ the confidence interval is

$$\left(\frac{14}{37} - \frac{7}{26} - 2.326 \times \sqrt{ \frac{14 \times (37 - 14)}{37^3} + \frac{7 \times (26 - 7)}{26^3} } , 1 \right)$$

$$= (-0.165, 1).$$

(d) With the pooled probability estimate

$$\hat{p} = \frac{x + y}{n + m} = \frac{14 + 7}{37 + 26} = 0.333$$

the test statistic is

$$z = \frac{\hat{p}_A - \hat{p}_B}{\sqrt{ \hat{p}(1 - \hat{p}) \left(\frac{1}{n} + \frac{1}{m} \right) }} = \frac{\frac{14}{37} - \frac{7}{26}}{\sqrt{ 0.333 \times (1 - 0.333) \times \left(\frac{1}{37} + \frac{1}{26} \right) }} = 0.905$$

and the p-value is

$$2 \times \Phi(-0.905) = 0.365.$$

10.2.2 (a) With $z_{0.005} = 2.576$ the confidence interval is

$$\left(\frac{261}{302} - \frac{401}{454} - 2.576 \times \sqrt{\frac{261 \times (302 - 261)}{302^3} + \frac{401 \times (454 - 401)}{454^3}} \, , \right.$$

$$\left. \frac{261}{302} - \frac{401}{454} + 2.576 \times \sqrt{\frac{261 \times (302 - 261)}{302^3} + \frac{401 \times (454 - 401)}{454^3}} \, \right)$$

$$= (-0.083, 0.045).$$

(b) With $z_{0.05} = 1.645$ the confidence interval is

$$\left(\frac{261}{302} - \frac{401}{454} - 1.645 \times \sqrt{\frac{261 \times (302 - 261)}{302^3} + \frac{401 \times (454 - 401)}{454^3}} \, , \right.$$

$$\left. \frac{261}{302} - \frac{401}{454} + 1.645 \times \sqrt{\frac{261 \times (302 - 261)}{302^3} + \frac{401 \times (454 - 401)}{454^3}} \, \right)$$

$$= (-0.060, 0.022).$$

(c) With $z_{0.05} = 1.645$ the confidence interval is

$$\left(-1 \, , \frac{261}{302} - \frac{401}{454} + 1.645 \times \sqrt{\frac{261 \times (302 - 261)}{302^3} + \frac{401 \times (454 - 401)}{454^3}} \, \right)$$

$$= (-1, 0.022).$$

(d) With the pooled probability estimate

$$\hat{p} = \frac{x + y}{n + m} = \frac{261 + 401}{302 + 454} = 0.876$$

the test statistic is

$$z = \frac{\hat{p}_A - \hat{p}_B}{\sqrt{\hat{p}(1 - \hat{p}) \left(\frac{1}{n} + \frac{1}{m} \right)}} = \frac{\frac{261}{302} - \frac{401}{454}}{\sqrt{0.876 \times (1 - 0.876) \times \left(\frac{1}{302} + \frac{1}{454} \right)}} = -0.776$$

and the p-value is

$$2 \times \Phi(-0.776) = 0.438.$$

10.2.3 (a) With $z_{0.005} = 2.576$ the confidence interval is

$$\left(\frac{35}{44} - \frac{36}{52} - 2.576 \times \sqrt{\frac{35 \times (44 - 35)}{44^3} + \frac{36 \times (52 - 36)}{52^3}} \, , \right.$$

$$\frac{35}{44} - \frac{36}{52} + 2.576 \times \sqrt{\frac{35 \times (44 - 35)}{44^3} + \frac{36 \times (52 - 36)}{52^3}}\,\Bigg)$$

$$= (-0.124, 0.331).$$

(b) With the pooled probability estimate

$$\hat{p} = \frac{x + y}{n + m} = \frac{35 + 36}{44 + 52} = 0.740$$

the test statistic is

$$z = \frac{\hat{p}_A - \hat{p}_B}{\sqrt{\hat{p}(1 - \hat{p})\left(\frac{1}{n} + \frac{1}{m}\right)}} = \frac{\frac{35}{44} - \frac{36}{52}}{\sqrt{0.740 \times (1 - 0.740) \times \left(\frac{1}{44} + \frac{1}{52}\right)}} = 1.147$$

and the p-value is

$$2 \times \Phi(-1.147) = 0.251.$$

There is *not* sufficient evidence to conclude that one radar system is any better than the other radar system.

10.2.4   (a) With $z_{0.005} = 2.576$ the confidence interval is

$$\left(\frac{4}{50} - \frac{10}{50} - 2.576 \times \sqrt{\frac{4 \times (50 - 4)}{50^3} + \frac{10 \times (50 - 10)}{50^3}}\,,\right.$$

$$\left.\frac{4}{50} - \frac{10}{50} + 2.576 \times \sqrt{\frac{4 \times (50 - 4)}{50^3} + \frac{10 \times (50 - 10)}{50^3}}\,\right)$$

$$= (-0.296, 0.056).$$

(b) With the pooled probability estimate

$$\hat{p} = \frac{x + y}{n + m} = \frac{4 + 10}{50 + 50} = 0.14$$

the test statistic is

$$z = \frac{\hat{p}_A - \hat{p}_B}{\sqrt{\hat{p}(1 - \hat{p})\left(\frac{1}{n} + \frac{1}{m}\right)}} = \frac{\frac{4}{50} - \frac{10}{50}}{\sqrt{0.14 \times (1 - 0.14) \times \left(\frac{1}{50} + \frac{1}{50}\right)}} = -1.729$$

and the p-value is

$$2 \times \Phi(-1.729) = 0.084.$$

(c) In this case the confidence interval is

$$\left(\frac{40}{500} - \frac{100}{500} - 2.576 \times \sqrt{\frac{40 \times (500 - 40)}{500^3} + \frac{100 \times (500 - 100)}{500^3}}, \right.$$

$$\left. \frac{40}{500} - \frac{100}{500} + 2.576 \times \sqrt{\frac{40 \times (500 - 40)}{500^3} + \frac{100 \times (500 - 100)}{500^3}} \right)$$

$$= (-0.176, -0.064).$$

With the pooled probability estimate

$$\hat{p} = \frac{x + y}{n + m} = \frac{40 + 100}{500 + 500} = 0.14$$

the test statistic is

$$z = \frac{\hat{p}_A - \hat{p}_B}{\sqrt{\hat{p}(1 - \hat{p})\left(\frac{1}{n} + \frac{1}{m}\right)}} = \frac{\frac{40}{500} - \frac{100}{500}}{\sqrt{0.14 \times (1 - 0.14) \times \left(\frac{1}{500} + \frac{1}{500}\right)}} = -5.468$$

and the p-value is

$$2 \times \Phi(-5.468) = 0.000.$$

10.2.5   Let p_A be the probability of crystallization within 24 hours *without* seed crystals and let p_B be the probability of crystallization within 24 hours *with* seed crystals.

With $z_{0.05} = 1.645$ a 95% upper confidence bound for $p_A - p_B$ is

$$\left(-1, \; \frac{27}{60} - \frac{36}{60} + 1.645 \times \sqrt{\frac{27 \times (60 - 27)}{60^3} + \frac{36 \times (60 - 36)}{60^3}} \right)$$

$$= (-1, -0.002).$$

Consider the hypothesis testing problem

$$H_0 : p_A \geq p_B \quad \text{versus} \quad H_A : p_A < p_B$$

where the alternative hypothesis states that the presence of seed crystals increases the probability of crystallization within 24 hours.

With the pooled probability estimate

$$\hat{p} = \frac{x + y}{n + m} = \frac{27 + 36}{60 + 60} = 0.525$$

the test statistic is

$$z = \frac{\hat{p}_A - \hat{p}_B}{\sqrt{\hat{p}(1 - \hat{p})\left(\frac{1}{n} + \frac{1}{m}\right)}} = \frac{\frac{27}{60} - \frac{36}{60}}{\sqrt{0.525 \times (1 - 0.525) \times \left(\frac{1}{60} + \frac{1}{60}\right)}} = -1.645$$

and the p-value is

$$\Phi(-1.645) = 0.050.$$

There is some evidence that the presence of seed crystals increases the probability of crystallization within 24 hours but it is not overwhelming.

10.2.6   Let p_A be the probability of an improved condition with the standard drug and let p_B be the probability of an improved condition with the new drug.

With $z_{0.05} = 1.645$ a 95% upper confidence bound for $p_A - p_B$ is

$$\left(-1 \, , \, \frac{72}{100} - \frac{83}{100} + 1.645 \times \sqrt{\frac{72 \times (100 - 72)}{100^3} + \frac{83 \times (100 - 83)}{100^3}} \right)$$

$$= (-1, -0.014).$$

Consider the hypothesis testing problem

$$H_0 : p_A \geq p_B \quad \text{versus} \quad H_A : p_A < p_B$$

where the alternative hypothesis states that the new drug increases the probability of an improved condition.

With the pooled probability estimate

$$\hat{p} = \frac{x + y}{n + m} = \frac{72 + 83}{100 + 100} = 0.775$$

the test statistic is

$$z = \frac{\hat{p}_A - \hat{p}_B}{\sqrt{\hat{p}(1 - \hat{p})\left(\frac{1}{n} + \frac{1}{m}\right)}} = \frac{\frac{72}{100} - \frac{83}{100}}{\sqrt{0.775 \times (1 - 0.775) \times \left(\frac{1}{100} + \frac{1}{100}\right)}} = -1.863$$

and the p-value is

$$\Phi(-1.863) = 0.031.$$

There is some evidence that the new drug increases the probability of an improved condition but it is not overwhelming.

10.2.7   Let p_A be the probability that a television set from production line A does not meet the quality standards and let p_B be the probability that a television set from production line B does not meet the quality standards.

With $z_{0.025} = 1.960$ a 95% two-sided confidence interval for $p_A - p_B$ is

$$\left(\frac{23}{1{,}128} - \frac{24}{962} - 1.960 \times \sqrt{\frac{23 \times (1{,}128 - 23)}{1{,}128^3} + \frac{24 \times (962 - 24)}{962^3}} \right. ,$$

$$\frac{23}{1,128} - \frac{24}{962} + 1.960 \times \sqrt{\frac{23 \times (1,128 - 23)}{1,128^3} + \frac{24 \times (962 - 24)}{962^3}}\Bigg)$$

$$= \ (-0.017, 0.008).$$

Consider the hypothesis testing problem

$$H_0 : p_A = p_B \quad \text{versus} \quad H_A : p_A \neq p_B$$

where the alternative hypothesis states that there is a difference in the operating standards of the two production lines.

With the pooled probability estimate

$$\hat{p} \ = \ \frac{x + y}{n + m} \ = \ \frac{23 + 24}{1,128 + 962} \ = \ 0.022$$

the test statistic is

$$z \ = \ \frac{\hat{p}_A - \hat{p}_B}{\sqrt{\hat{p}(1 - \hat{p})\left(\frac{1}{n} + \frac{1}{m}\right)}} \ = \ \frac{\frac{23}{1,128} - \frac{24}{962}}{\sqrt{0.022 \times (1 - 0.022) \times \left(\frac{1}{1,128} + \frac{1}{962}\right)}} \ = \ -0.708$$

and the p-value is

$$2 \times \Phi(-0.708) \ = \ 0.479.$$

There is *not* sufficient evidence to conclude that there is a difference in the operating standards of the two production lines.

10.2.8   Let p_A be the probability that the armor plating of a tank is pierced by a standard artillery shell and let p_B be the probability that the armor plating of a tank is pierced by a new artillery shell.

With $z_{0.05} = 1.645$ a 95% upper confidence bound for $p_A - p_B$ is

$$\left(-1 \ , \ \frac{73}{120} - \frac{80}{120} + 1.645 \times \sqrt{\frac{73 \times (120 - 73)}{120^3} + \frac{80 \times (120 - 80)}{120^3}}\right)$$

$$= \ (-1, 0.044).$$

Consider the hypothesis testing problem

$$H_0 : p_A \geq p_B \quad \text{versus} \quad H_A : p_A < p_B$$

where the alternative hypothesis states that the new artillery shell increases the probability that the armor plating of a tank is pierced.

With the pooled probability estimate

$$\hat{p} = \frac{x+y}{n+m} = \frac{73+80}{120+120} = 0.6375$$

the test statistic is

$$z = \frac{\hat{p}_A - \hat{p}_B}{\sqrt{\hat{p}(1-\hat{p})\left(\frac{1}{n}+\frac{1}{m}\right)}} = \frac{\frac{73}{120} - \frac{80}{120}}{\sqrt{0.6375 \times (1 - 0.6375) \times \left(\frac{1}{120}+\frac{1}{120}\right)}} = -0.940$$

and the p-value is

$$\Phi(-0.940) = 0.174.$$

There is *not* sufficient evidence to conclude that the new artillery shell increases the probability that the armor plating of a tank is pierced.

10.2.9   Let p_A be the probability that a computer chip from supplier A is defective and let p_B be the probability that a computer chip from supplier B is defective.

With $z_{0.025} = 1.960$ a 95% two-sided confidence interval for $p_A - p_B$ is

$$\left(\frac{8}{200} - \frac{13}{250} - 1.960 \times \sqrt{\frac{8 \times (200-8)}{200^3} + \frac{13 \times (250-13)}{250^3}} \ , \right.$$

$$\left. \frac{8}{200} - \frac{13}{250} + 1.960 \times \sqrt{\frac{8 \times (200-8)}{200^3} + \frac{13 \times (250-13)}{250^3}} \right)$$

$$= (-0.051, 0.027).$$

Consider the hypothesis testing problem

$$H_0 : p_A = p_B \quad \text{versus} \quad H_A : p_A \neq p_B$$

where the alternative hypothesis states that there is a difference in the quality of the computer chips from the two suppliers.

With the pooled probability estimate

$$\hat{p} = \frac{x+y}{n+m} = \frac{8+13}{200+250} = 0.047$$

the test statistic is

$$z = \frac{\hat{p}_A - \hat{p}_B}{\sqrt{\hat{p}(1-\hat{p})\left(\frac{1}{n}+\frac{1}{m}\right)}} = \frac{\frac{8}{200} - \frac{13}{250}}{\sqrt{0.047 \times (1 - 0.047) \times \left(\frac{1}{200}+\frac{1}{250}\right)}} = -0.600$$

and the p-value is

$$2 \times \Phi(-0.600) = 0.549.$$

There is *not* sufficient evidence to conclude that there is a difference in the quality of the computer chips from the two suppliers.

10.2.10 Let p_A be the probability of an error in an application processed during the first two weeks and let p_B be the probability of an error in an application processed after the first two weeks.

With $z_{0.05} = 1.645$ a 95% lower confidence bound for $p_A - p_B$ is

$$\left(\frac{17}{85} - \frac{16}{132} - 1.645 \times \sqrt{ \frac{17 \times (85 - 17)}{85^3} + \frac{16 \times (132 - 16)}{132^3} } , \ 1 \right)$$

$$= (-0.007, 1).$$

Consider the hypothesis testing problem

$$H_0 : p_A \leq p_B \quad \text{versus} \quad H_A : p_A > p_B$$

where the alternative hypothesis states that the probability of an error in the processing of an application is larger during the first two weeks.

With the pooled probability estimate

$$\hat{p} = \frac{x + y}{n + m} = \frac{17 + 16}{85 + 132} = 0.152$$

the test statistic is

$$z = \frac{\hat{p}_A - \hat{p}_B}{\sqrt{\hat{p}(1 - \hat{p}) \left(\frac{1}{n} + \frac{1}{m} \right)}} = \frac{\frac{17}{85} - \frac{16}{132}}{\sqrt{0.152 \times (1 - 0.152) \times \left(\frac{1}{85} + \frac{1}{132} \right)}} = 1.578$$

and the p-value is

$$1 - \Phi(1.578) = 0.057.$$

There is some evidence that the probability of an error in the processing of an application is larger during the first two weeks but it is not overwhelming.

10.2.11 With the pooled probability estimate

$$\hat{p} = \frac{x + y}{n + m} = \frac{11 + 7}{18 + 18} = 0.5$$

the test statistic is

$$z = \frac{\hat{p}_A - \hat{p}_B}{\sqrt{\hat{p}(1 - \hat{p})\left(\frac{1}{n} + \frac{1}{m}\right)}} = \frac{\frac{11}{18} - \frac{7}{18}}{\sqrt{0.5 \times (1 - 0.5) \times \left(\frac{1}{18} + \frac{1}{18}\right)}} = 1.333$$

and the p-value is

$$2 \times \Phi(-1.333) = 0.183.$$

The null hypothesis $(H_0 : p_A = p_B)$ is accepted for a size $\alpha = 0.05$ test. There is *not* sufficient evidence to conclude that there is any difference in the prediction abilities of the two commentators.

10.2.12   Let p_A be the probability that a card is left in the machine with the original instructions and let p_B be the probability that a card is left in the machine with the new instructions.

With $z_{0.05} = 1.645$ a 95% lower confidence bound for $p_A - p_B$ is

$$\left(\frac{22}{542} - \frac{16}{526} - 1.645 \times \sqrt{\frac{22 \times (542 - 22)}{542^3} + \frac{16 \times (526 - 16)}{526^3}} \, , \, 1 \right)$$

$$= (-0.008, 1).$$

Consider the hypothesis testing problem

$$H_0 : p_A \leq p_B \quad \text{versus} \quad H_A : p_A > p_B$$

where the alternative hypothesis states that the probability of leaving a card in the machine is smaller with the new set of instructions.

With the pooled probability estimate

$$\hat{p} = \frac{x + y}{n + m} = \frac{22 + 16}{542 + 526} = 0.0356$$

the test statistic is

$$z = \frac{\hat{p}_A - \hat{p}_B}{\sqrt{\hat{p}(1 - \hat{p})\left(\frac{1}{n} + \frac{1}{m}\right)}} = \frac{\frac{22}{542} - \frac{16}{526}}{\sqrt{0.0356 \times (1 - 0.0356) \times \left(\frac{1}{542} + \frac{1}{526}\right)}} = 0.900$$

and the p-value is

$$1 - \Phi(0.900) = 0.184.$$

There is *not* sufficient evidence to conclude that the probability of leaving a card in the machine is smaller with the new set of instructions.

10.3 Goodness of Fit Tests for One-way Contingency Tables

10.3.1   (a) The expected cell frequencies are $e_i = \frac{500}{6} = 83.33$.

(b) The Pearson chi-square statistic is

$$X^2 = \frac{(80-83.33)^2}{83.33} + \frac{(71-83.33)^2}{83.33} + \frac{(90-83.33)^2}{83.33} + \frac{(87-83.33)^2}{83.33}$$
$$+ \frac{(78-83.33)^2}{83.33} + \frac{(94-83.33)^2}{83.33} = 4.36.$$

(c) The liklihood ratio chi-square statistic is

$$G^2 = 2 \times \left(80\ln\left(\frac{80}{83.33}\right) + 71\ln\left(\frac{71}{83.33}\right) + 90\ln\left(\frac{90}{83.33}\right)\right.$$
$$\left. + 87\ln\left(\frac{87}{83.33}\right) + 78\ln\left(\frac{78}{83.33}\right) + 94\ln\left(\frac{94}{83.33}\right)\right) = 4.44.$$

(d) The p-values are

$$P(X \geq 4.36) = 0.499 \quad \text{and} \quad P(X \geq 4.44) = 0.488$$

where the random variable X has a chi-square distribution with $6 - 1 = 5$ degrees of freedom.

A size $\alpha = 0.01$ test of the null hypothesis that the die is fair is accepted.

(e) With $z_{0.05} = 1.645$ the confidence interval is

$$\left(\frac{94}{500} - \frac{1.645}{500} \times \sqrt{\frac{94 \times (500-94)}{500}}, \frac{94}{500} + \frac{1.645}{500} \times \sqrt{\frac{94 \times (500-94)}{500}}\right)$$

$$= (0.159, 0.217).$$

10.3.2   The expected cell frequencies are

1	2	3	4	5	6	7	8	9	≥ 10
50.00	41.67	34.72	28.94	24.11	20.09	16.74	13.95	11.62	58.16

The Pearson chi-square statistic is $X^2 = 10.33$.

The p-value is

$$P(X \geq 10.33) = 0.324$$

where the random variable X has a chi-square distribution with $10 - 1 = 9$ degrees of freedom.

The geometric distribution with $p = 1/6$ is plausible.

10.3.3   (a) The expected cell frequencies are

$e_1 = 221 \times \frac{4}{7} = 126.29$,

$e_2 = 221 \times \frac{2}{7} = 63.14$, and

$e_3 = 221 \times \frac{1}{7} = 31.57$.

The Pearson chi-square statistic is

$$X^2 = \frac{(113 - 126.29)^2}{126.29} + \frac{(82 - 63.14)^2}{63.14} + \frac{(26 - 31.57)^2}{31.57} = 8.01.$$

The p-value is

$$P(X \geq 8.01) = 0.018$$

where the random variable X has a chi-square distribution with $3 - 1 = 2$ degrees of freedom.

There is a strong suggestion that the supposition is not plausible although the evidence is not completely overwhelming.

(b) With $z_{0.005} = 2.576$ the confidence interval is

$$\left(\frac{113}{221} - \frac{2.576}{221} \times \sqrt{\frac{113 \times (221 - 113)}{221}}, \frac{113}{221} + \frac{2.576}{221} \times \sqrt{\frac{113 \times (221 - 113)}{221}} \right)$$

$$= (0.425, 0.598).$$

10.3.4   The expected cell frequencies are

$e_1 = 964 \times 0.14 = 134.96$,

$e_2 = 964 \times 0.22 = 212.08$,

$e_3 = 964 \times 0.35 = 337.40$,

$e_4 = 964 \times 0.16 = 154.24$, and

$e_5 = 964 \times 0.13 = 125.32$.

The Pearson chi-square statistic is $X^2 = 14.6$.

The p-value is

$$P(X \geq 14.6) = 0.006$$

where the random variable X has a chi-square distribution with $5 - 1 = 4$ degrees of freedom.

There is sufficient evidence to conclude that the jurors have not been selected randomly.

10.3.5   (a) The expected cell frequencies are

$e_1 = 126 \times 0.5 = 63.0$,

$e_2 = 126 \times 0.4 = 50.4$, and

$e_3 = 126 \times 0.1 = 12.6$.

The likelihood ratio chi-square statistic is

$$G^2 = 2 \times \left(56 \ln \left(\frac{56}{63.0} \right) + 51 \ln \left(\frac{51}{50.4} \right) + 19 \ln \left(\frac{19}{12.6} \right) \right) = 3.62.$$

The p-value is

$$P(X \geq 3.62) = 0.164$$

where the random variable X has a chi-square distribution with $3 - 1 = 2$ degrees of freedom.

These probability values are plausible.

(b) With $z_{0.025} = 1.960$ the confidence interval is

$$\left(\frac{56}{126} - \frac{1.960}{126} \times \sqrt{\frac{56 \times (126 - 56)}{126}}, \frac{56}{126} + \frac{1.960}{126} \times \sqrt{\frac{56 \times (126 - 56)}{126}} \right)$$

$$= (0.358, 0.531).$$

10.3.6 If the three soft drink formulations are equally likely then the expected cell frequencies are $e_i = 600 \times \frac{1}{3} = 200$.

The Pearson chi-square statistic is

$$X^2 = \frac{(225 - 200)^2}{200} + \frac{(223 - 200)^2}{200} + \frac{(152 - 200)^2}{200} = 17.29.$$

The p-value is

$$P(X \geq 17.29) = 0.0002$$

where the random variable X has a chi-square distribution with $3 - 1 = 2$ degrees of freedom.

It is not plausible that the three soft drink formulations are equally likely.

10.3.7 The first two cells should be pooled so that there are 13 cells altogether.

The Pearson chi-square statistic is $X^2 = 92.9$ and the p-value is

$$P(X \geq 92.9) = 0.0000$$

where the random variable X has a chi-square distribution with $13 - 1 = 12$ degrees of freedom.

It is not reasonable to model the number of arrivals with a Poisson distribution with mean $\lambda = 7$.

10.3.8   A Poisson distribution with mean $\lambda = \bar{x} = 4.49$ can be considered. The first two cells should be pooled and the last two cells should be pooled so that there are 9 cells altogether.

The Pearson chi-square statistic is $X^2 = 8.3$ and the p-value is

$$P(X \geq 8.3) = 0.307$$

where the random variable X has a chi-square distribution with $9 - 2 = 7$ degrees of freedom.

It is reasonable to model the number of radioactive particles emitted with a Poisson distribution.

10.3.9   If the pearl oyster diameters have a uniform distribution then the expected cell frequencies are

$e_1 = 1490 \times 0.1 = 149,$

$e_2 = 1490 \times 0.2 = 298,$

$e_3 = 1490 \times 0.2 = 298,$ and

$e_4 = 1490 \times 0.5 = 745.$

The Pearson chi-square statistic is

$$X^2 = \frac{(161 - 149)^2}{149} + \frac{(289 - 298)^2}{298} + \frac{(314 - 298)^2}{298} + \frac{(726 - 745)^2}{745} = 2.58.$$

The p-value is

$$P(X \geq 2.58) = 0.461$$

where the random variable X has a chi-square distribution with $4 - 1 = 3$ degrees of freedom.

It is plausible that the pearl oyster diameters have a uniform distribution between 0 and 10 mm.

10.4 Testing for Independence in Two-way Contingency Tables

10.4.1   (a) The expected cell frequencies are

	acceptable	defective
Supplier A	186.25	13.75
Supplier B	186.25	13.75
Supplier C	186.25	13.75
Supplier D	186.25	13.75

      (b) The Pearson chi-square statistic is $X^2 = 7.087$.

      (c) The likelihood ratio chi-square statistic is $G^2 = 6.889$.

      (d) The p-values are

$$P(X \geq 7.087) = 0.069 \quad \text{and} \quad P(X \geq 6.889) = 0.076$$

where the random variable X has a chi-square distribution with $(4-1) \times (2-1) = 3$ degrees of freedom.

      (e) The null hypothesis that the defective rates are identical for the four suppliers is accepted at size $\alpha = 0.05$.

      (f) With $z_{0.025} = 1.960$ the confidence interval is

$$\left(\frac{10}{200} - \frac{1.960}{200} \times \sqrt{\frac{10 \times (200 - 10)}{200}}, \frac{10}{200} + \frac{1.960}{200} \times \sqrt{\frac{10 \times (200 - 10)}{200}} \right)$$

$$= (0.020, 0.080).$$

      (g) With $z_{0.025} = 1.960$ the confidence interval is

$$\left(\frac{15}{200} - \frac{21}{200} - 1.960 \times \sqrt{\frac{15 \times (200 - 15)}{200^3} + \frac{21 \times (200 - 21)}{200^3}}, \right.$$

$$\left. \frac{15}{200} - \frac{21}{200} + 1.960 \times \sqrt{\frac{15 \times (200 - 15)}{200^3} + \frac{21 \times (200 - 21)}{200^3}} \right)$$

$$= (-0.086, 0.026).$$

10.4.2   The expected cell frequencies are

	dead	slow growth	medium growth	strong growth
no fertilizer	57.89	93.84	172.09	163.18
fertilizer I	61.22	99.23	181.98	172.56
no fertilizer II	62.89	101.93	186.93	177.25

The Pearson chi-square statistic is $X^2 = 13.66$.

The p-value is

$$P(X \geq 13.66) = 0.034$$

where the random variable X has a chi-square distribution with $(3-1) \times (4-1) = 6$ degrees of freedom.

There is a fairly strong suggestion that the seedlings growth pattern is different for the different growing conditions although the evidence is not overwhelming.

10.4.3   The expected cell frequencies are

	formulation I	formulation II	formulation III
10-25	75.00	74.33	50.67
26-50	75.00	74.33	50.67
≥ 51	75.00	74.33	50.67

The Pearson chi-square statistic is $X^2 = 6.11$.

The p-value is

$$P(X \geq 6.11) = 0.191$$

where the random variable X has a chi-square distribution with $(3-1) \times (3-1) = 4$ degrees of freedom.

There is *not* sufficient evidence to conclude that the preferences for the different formulations change with age.

10.4.4   (a) The expected cell frequencies are

	pass	fail
line 1	166.2	13.8
line 2	166.2	13.8
line 3	166.2	13.8
line 4	166.2	13.8
line 5	166.2	13.8

The Pearson chi-square statistic is $X^2 = 13.72$.

The p-value is

$$P(X \geq 13.72) = 0.008$$

where the random variable X has a chi-square distribution with $(5-1) \times (2-1) = 4$ degrees of freedom.

There is sufficient evidence to conclude that the pass rates are different for the five production lines.

(b) With $z_{0.025} = 1.960$ the confidence interval is

$$\left(\frac{11}{180} - \frac{15}{180} - 1.960 \times \sqrt{\frac{11 \times (180-11)}{180^3} + \frac{15 \times (180-15)}{180^3}} \, , \right.$$

$$\left. \frac{11}{180} - \frac{15}{180} + 1.960 \times \sqrt{\frac{11 \times (180-11)}{180^3} + \frac{15 \times (180-15)}{180^3}} \right)$$

$$= (-0.076, 0.031).$$

10.4.5   The expected cell frequencies are

	completely satisfied	somewhat satisfied	not satisfied
Technician 1	71.50	22.36	4.14
Technician 2	83.90	26.24	4.86
Technician 3	45.96	14.37	2.66
Technician 4	57.64	18.03	3.34

The Pearson chi-square statistic is $X^2 = 32.11$.

The p-value is

$$P(X \geq 32.11) = 0.000$$

where the random variable X has a chi-square distribution with $(4-1) \times (3-1) = 6$ degrees of freedom.

There is sufficient evidence to conclude that some technicians are better than others in satisfying their customers.

Note: In this analysis 4 of the cells have expected values less than 5 and it may be preferable to pool together the categories "somewhat satisfied" and "not satisfied". In this case the Pearson chi-square statistic is $X^2 = 31.07$ and comparison with a chi-square distribution with 3 degrees of freedom again gives a p-value of 0.000. The conclusion remains the same.

10.4.7   (a) The expected cell frequencies are

	less than one week	more than one week
standard drug	88.63	64.37
new drug	79.37	57.63

The Pearson chi-square statistic is $X^2 = 15.71$.

The p-value is

$$P(X \geq 15.71) \;=\; 0.0000$$

where the random variable X has a chi-square distribution with $(2-1) \times (2-1) = 1$ degree of freedom.

There is sufficient evidence to conclude that $p_s \neq p_n$.

(b) With $z_{0.005} = 2.576$ the confidence interval is

$$\left(\frac{72}{153} - \frac{96}{137} - 2.576 \times \sqrt{\frac{72 \times (153 - 72)}{153^3} + \frac{96 \times (137 - 96)}{137^3}} \; , \right.$$

$$\left. \frac{72}{153} - \frac{96}{137} + 2.576 \times \sqrt{\frac{72 \times (153 - 72)}{153^3} + \frac{96 \times (137 - 96)}{137^3}} \right)$$

$$= \; (-0.375, -0.085).$$

10.5   Supplementary Problems

10.5.1   With $z_{0.025} = 1.960$ the confidence interval is

$$\left(\frac{27}{60} - \frac{1.960}{60} \times \sqrt{\frac{27 \times (60 - 27)}{60}}, \frac{27}{60} + \frac{1.960}{60} \times \sqrt{\frac{27 \times (60 - 27)}{60}} \right)$$

$$= (0.324, 0.576).$$

10.5.2   Let p be the probability that a bag of flour is underweight and consider the hypothesis testing problem

$$H_0 : p \leq \frac{1}{40} = 0.025 \quad \text{versus} \quad H_A : p > \frac{1}{40} = 0.025$$

where the alternative hypothesis states that the consumer watchdog organization can take legal action.

The statistic for the normal approximation of the p-value is

$$z = \frac{x - np_0}{\sqrt{np_0(1 - p_0)}} = \frac{18 - (500 \times 0.025)}{\sqrt{500 \times 0.025 \times (1 - 0.025)}} = 1.575$$

and the p-value is

$$1 - \Phi(1.575) = 0.058.$$

There is a fairly strong suggestion that the proportion of underweight bags is more than 1 in 40 although the evidence is not overwhelming.

10.5.3   Let p be the proportion of customers who request the credit card. With $z_{0.005} = 2.576$ a 95% two-sided confidence interval for p is

$$\left(\frac{384}{5,000} - \frac{2.576}{5,000} \times \sqrt{\frac{384 \times (5,000 - 384)}{5,000}}, \frac{384}{5,000} + \frac{2.576}{5,000} \times \sqrt{\frac{384 \times (5,000 - 384)}{5,000}} \right)$$

$$= (0.0671, 0.0865).$$

The number of customers out of 1,000,000 who request the credit card can then be estimated as being between 67,100 and 86,500.

10.5.4   Let p_A be the probability that an operation performed in the morning is a total success and let p_B be the probability that an operation performed in the afternoon is a total success.

With $z_{0.05} = 1.645$ a 95% lower confidence bound for $p_A - p_B$ is

$$\left(\frac{443}{564} - \frac{388}{545} - 1.645 \times \sqrt{\frac{443 \times (564 - 443)}{564^3} + \frac{388 \times (545 - 388)}{545^3}} \, , \, 1 \right)$$

$$= (0.031, 1).$$

Consider the hypothesis testing problem

$$H_0 : p_A \leq p_B \quad \text{versus} \quad H_A : p_A > p_B$$

where the alternative hypothesis states that the probability that an operation is a total success is smaller in the afternoon than in the morning.

With the pooled probability estimate

$$\hat{p} = \frac{x + y}{n + m} = \frac{443 + 388}{564 + 545} = 0.749$$

the test statistic is

$$z = \frac{\hat{p}_A - \hat{p}_B}{\sqrt{\hat{p}(1 - \hat{p}) \left(\frac{1}{n} + \frac{1}{m} \right)}} = \frac{\frac{443}{564} - \frac{388}{545}}{\sqrt{0.749 \times (1 - 0.749) \times \left(\frac{1}{564} + \frac{1}{545} \right)}} = 2.822$$

and the p-value is

$$1 - \Phi(2.822) = 0.002.$$

There is sufficient evidence to conclude that the probability that an operation is a total success is smaller in the afternoon than in the morning.

10.5.5   Let p_A be the probability that a householder with an income above \$60,000 supports the tax increase and let p_B be the probability that a householder with an income below \$60,000 supports the tax increase.

With $z_{0.025} = 1.960$ a 95% two-sided confidence interval for $p_A - p_B$ is

$$\left(\frac{32}{106} - \frac{106}{221} - 1.960 \times \sqrt{\frac{32 \times (106 - 32)}{106^3} + \frac{106 \times (221 - 106)}{221^3}} \, , \right.$$

$$\left. \frac{32}{106} - \frac{106}{221} + 1.960 \times \sqrt{\frac{32 \times (106 - 32)}{106^3} + \frac{106 \times (221 - 106)}{221^3}} \right)$$

$$= (-0.287, -0.068).$$

Consider the hypothesis testing problem

$$H_0 : p_A = p_B \quad \text{versus} \quad H_A : p_A \neq p_B$$

where the alternative hypothesis states that the support for the tax increase does depend upon the householder's income.

With the pooled probability estimate

$$\hat{p} = \frac{x + y}{n + m} = \frac{32 + 106}{106 + 221} = 0.422$$

the test statistic is

$$z = \frac{\hat{p}_A - \hat{p}_B}{\sqrt{\hat{p}(1 - \hat{p}) \left(\frac{1}{n} + \frac{1}{m} \right)}} = \frac{\frac{32}{106} - \frac{106}{221}}{\sqrt{0.422 \times (1 - 0.422) \times \left(\frac{1}{106} + \frac{1}{221} \right)}} = -3.05$$

and the p-value is

$$2 \times \Phi(-3.05) = 0.002.$$

There is sufficient evidence to conclude that the support for the tax increase does depend upon the householder's income.

10.5.6   The expected cell frequencies are

$e_1 = 619 \times 0.1 = 61.9,$

$e_2 = 619 \times 0.8 = 495.2,$ and

$e_3 = 619 \times 0.1 = 61.9.$

The Pearson chi-square statistic is

$$X^2 = \frac{(61 - 61.9)^2}{61.9} + \frac{(486 - 495.2)^2}{495.2} + \frac{(72 - 61.9)^2}{61.9} = 1.83.$$

The p-value is

$$P(X \geq 3.62) = 0.400$$

where the random variable X has a chi-square distribution with $3 - 1 = 2$ degrees of freedom.

These probability values are plausible.

10.5.7   A Poisson distribution with mean $\lambda = \bar{x} = 2.95$ can be considered. The last two cells can be pooled so that there are 8 cells altogether.

The Pearson chi-square statistic is $X^2 = 13.1$ and the p-value is

$$P(X \geq 13.1) \; = \; 0.041$$

where the random variable X has a chi-square distribution with $8 - 2 = 6$ degrees of freedom.

There is some evidence that a Poisson distribution is not appropriate although the evidence is not overwhelming.

10.5.8   If the random numbers have a uniform distribution then the expected cell frequencies are $e_i = 1,000$.

The Pearson chi-square statistic is $X^2 = 9.07$ and the p-value is

$$P(X \geq 9.07) \; = \; 0.431$$

where the random variable X has a chi-square distribution with $10 - 1 = 9$ degrees of freedom.

There is no evidence that the random number generator is not operating correctly.

10.5.10   The expected cell frequencies are

	A	B	C
this year	112.58	78.18	30.23
last year	211.42	146.82	56.77

The Pearson chi-square statistic is $X^2 = 1.20$.

The p-value is

$$P(X \geq 1.20) \; = \; 0.549$$

where the random variable X has a chi-square distribution with $(2-1) \times (3-1) = 2$ degrees of freedom.

There is *not* sufficient evidence to conclude that there has been a change in preferences for the three types of tire between the two years.

10.5.11   The expected cell frequencies are

	completely destroyed	partially destroyed	missed
team alpha	19.56	17.81	6.63
team beta	22.22	20.24	7.54
team gamma	14.22	12.95	4.83

The Pearson chi-square statistic is $X^2 = 5.66$.

The p-value is

$$P(X \geq 5.66) = 0.226$$

where the random variable X has a chi-square distribution with $(3-1) \times (3-1) = 4$ degrees of freedom.

There is *not* sufficient evidence to conclude that the three teams are not equally skillful.

10.5.12　The expected cell frequencies are

	Yes	No
engineering	72.09	70.91
arts & sciences	49.91	49.09

The Pearson chi-square statistic is $X^2 = 4.28$.

The p-value is

$$P(X \geq 4.28) = 0.039$$

where the random variable X has a chi-square distribution with $(2-1) \times (2-1) = 1$ degree of freedom.

There is a fairly strong suggestion that the opinions differ between the two colleges but the evidence is not overwhelming.

Chapter 11

The Analysis of Variance

11.1 One Factor Analysis of Variance

11.1.1 (a) $P(X \geq 4.2) = 0.0177$.

(b) $P(X \geq 2.3) = 0.0530$.

(c) $P(X \geq 31.7) \leq 0.0001$.

(d) $P(X \geq 9.3) = 0.0019$.

(e) $P(X \geq 8.4) \leq 0.0001$.

11.1.2

Source	df	SS	MS	F	p-value
Treatments	5	557.0	111.4	5.547	0.0017
Error	23	461.9	20.08		
Total	28	1018.9			

11.1.3

Source	df	SS	MS	F	p-value
Treatments	7	126.95	18.136	5.01	0.0016
Error	22	79.64	3.62		
Total	29	206.59			

11.1.4

Source	df	SS	MS	F	p-value
Treatments	6	7.66	1.28	0.78	0.59
Error	77	125.51	1.63		
Total	83	133.18			

11.1.5

Source	df	SS	MS	F	p-value
Treatments	3	162.19	54.06	6.69	0.001
Error	40	323.34	8.08		
Total	43	485.53			

11.1.6

Source	df	SS	MS	F	p-value
Treatments	2	48.14	24.07	2.79	0.071
Error	52	449.25	8.64		
Total	54	497.39			

11.1.7

Source	df	SS	MS	F	p-value
Treatments	3	0.0079	0.0026	1.65	0.189
Error	52	0.0829	0.0016		
Total	55	0.0908			

11.1.8 (a)
$$\mu_1 - \mu_2 \in \left(48.05 - 44.74 - \frac{\sqrt{4.96} \times 3.49}{\sqrt{11}}, 48.05 - 44.74 + \frac{\sqrt{4.96} \times 3.49}{\sqrt{11}} \right)$$

$$= (0.97, 5.65).$$

$$\mu_1 - \mu_3 \in \left(48.05 - 49.11 - \frac{\sqrt{4.96} \times 3.49}{\sqrt{11}}, 48.05 - 49.11 + \frac{\sqrt{4.96} \times 3.49}{\sqrt{11}} \right)$$

$$= (-3.40, 1.28).$$

$$\mu_2 - \mu_3 \in \left(44.74 - 49.11 - \frac{\sqrt{4.96} \times 3.49}{\sqrt{11}}, 44.74 - 49.11 + \frac{\sqrt{4.96} \times 3.49}{\sqrt{11}} \right)$$

$$= (-6.71, -2.03).$$

(c) The total sample size required from each factor level can be estimated as

$$n \geq \frac{4\, s^2\, q^2_{\alpha,k,\nu}}{L^2} = \frac{4 \times 4.96 \times 3.49^2}{2.0^2} = 60.4$$

so that an additional sample size of $61 - 11 = 50$ observations from each factor level can be recommended.

11.1.9 (a)
$$\mu_1 - \mu_2 \in \left(136.3 - 152.1 - \frac{\sqrt{15.95} \times 4.30}{\sqrt{6}}, 136.3 - 152.1 + \frac{\sqrt{15.95} \times 4.30}{\sqrt{6}} \right)$$

$$= (-22.8, -8.8).$$

$$\mu_1 - \mu_3 \in (3.6, 17.6).$$

$$\mu_1 - \mu_4 \in (-0.9, 13.1).$$

$$\mu_1 - \mu_5 \in (-13.0, 1.0).$$

$$\mu_1 - \mu_6 \in (1.3, 15.3).$$

$$\mu_2 - \mu_3 \in (19.4, 33.4).$$

$$\mu_2 - \mu_4 \in (14.9, 28.9).$$

$$\mu_2 - \mu_5 \in (2.8, 16.8).$$

$$\mu_2 - \mu_6 \in (17.1, 31.1).$$

$$\mu_3 - \mu_4 \in (-11.5, 2.5).$$

$$\mu_3 - \mu_5 \in (-23.6, -9.6).$$

$$\mu_3 - \mu_6 \in (-9.3, 4.7).$$

$$\mu_4 - \mu_5 \in (-19.1, -5.1).$$

$$\mu_4 - \mu_6 \in (-4.8, 9.2).$$

$$\mu_5 - \mu_6 \in (7.3, 21.3).$$

(c) The total sample size required from each factor level can be estimated as

$$n \geq \frac{4\, s^2\, q_{\alpha,k,\nu}^2}{L^2} = \frac{4 \times 15.95 \times 4.30^2}{10.0^2} = 11.8$$

so that an additional sample size of $12 - 6 = 6$ observations from each factor level can be recommended.

11.1.10 The p-value remains unchanged.

11.1.11 (a) $\bar{x}_{1.} = 5.633.$
$\bar{x}_{2.} = 5.567.$
$\bar{x}_{3.} = 4.778.$

(b) $\bar{x}_{..} = 5.326.$

(c) $SSTR = 4.076.$

(d) $\sum_{i=1}^{k} \sum_{j=1}^{n_i} x_{ij}^2 = 791.30.$

(e) $SST = 25.432.$

(f) $SSE = 21.356$.

(g)

Source	df	SS	MS	F	p-value
Treatments	2	4.076	2.038	2.29	0.123
Error	24	21.356	0.890		
Total	26	25.432			

(h)
$$\mu_1 - \mu_2 \in \left(5.633 - 5.567 - \frac{\sqrt{0.890} \times 3.53}{\sqrt{9}}, 5.633 - 5.567 + \frac{\sqrt{0.890} \times 3.53}{\sqrt{9}} \right)$$

$$= (-1.04, 1.18).$$

$$\mu_1 - \mu_3 \in \left(5.633 - 4.778 - \frac{\sqrt{0.890} \times 3.53}{\sqrt{9}}, 5.633 - 4.778 + \frac{\sqrt{0.890} \times 3.53}{\sqrt{9}} \right)$$

$$= (-0.25, 1.97).$$

$$\mu_2 - \mu_3 \in \left(5.567 - 4.778 - \frac{\sqrt{0.890} \times 3.53}{\sqrt{9}}, 5.567 - 4.778 + \frac{\sqrt{0.890} \times 3.53}{\sqrt{9}} \right)$$

$$= (-0.32, 1.90).$$

(j) The total sample size required from each factor level can be estimated as

$$n \geq \frac{4 \, s^2 \, q_{\alpha,k,\nu}^2}{L^2} = \frac{4 \times 0.890 \times 3.53^2}{1.0^2} = 44.4$$

so that an additional sample size of $45 - 9 = 36$ observations from each factor level can be recommended.

11.1.12   (a) $\bar{x}_{1.} = 10.560$.
$\bar{x}_{2.} = 15.150$.
$\bar{x}_{3.} = 17.700$.
$\bar{x}_{4.} = 11.567$.

(b) $\bar{x}_{..} = 14.127$.

(c) $SSTR = 364.75$.

(d) $\sum_{i=1}^{k} \sum_{j=1}^{n_i} x_{ij}^2 = 9,346.74$.

(e) $SST = 565.23$.

(f) $SSE = 200.47$.

(g)

Source	df	SS	MS	F	p-value
Treatments	3	364.75	121.58	24.26	0.000
Error	40	200.47	5.01		
Total	43	565.23			

(h) $\mu_1 - \mu_2 \in (-7.16, -2.02)$.

$\mu_1 - \mu_3 \in (-9.66, -4.62)$.

$\mu_1 - \mu_4 \in (-3.76, 1.75)$.

$\mu_2 - \mu_3 \in (-4.95, -0.15)$.

$\mu_2 - \mu_4 \in (0.94, 6.23)$.

$\mu_3 - \mu_4 \in (3.53, 8.74)$.

Note: In the remainder of this section confidence intervals for the pairwise differences of the factor level means are provided with an overall confidence level of 95%.

11.1.13

Source	df	SS	MS	F	p-value
Treatments	2	0.0085	0.0042	0.24	0.787
Error	87	1.5299	0.0176		
Total	89	1.5384			

$\mu_1 - \mu_2 \in (-0.08, 0.08)$.

$\mu_1 - \mu_3 \in (-0.06, 0.10)$.

$\mu_2 - \mu_3 \in (-0.06, 0.10)$.

There is *not* sufficient evidence to conclude that there is a difference between the three production lines.

11.1.14

Source	df	SS	MS	F	p-value
Treatments	2	278.0	139.0	85.4	0.000
Error	50	81.3	1.63		
Total	52	359.3			

$\mu_1 - \mu_2 \in (3.06, 5.16)$.

$\mu_1 - \mu_3 \in (4.11, 6.11)$.

$\mu_2 - \mu_3 \in (-0.08, 2.08)$.

There is sufficient evidence to conclude that Monday is slower than the other two days.

11.1.15

Source	df	SS	MS	F	p-value
Treatments	2	0.0278	0.0139	1.26	0.299
Error	30	0.3318	0.0111		
Total	32	0.3596			

$\mu_1 - \mu_2 \in (-0.15, 0.07)$.

$\mu_1 - \mu_3 \in (-0.08, 0.14)$.

$\mu_2 - \mu_3 \in (-0.04, 0.18)$.

There is *not* sufficient evidence to conclude that the radiation readings are affected by the background radiation levels.

11.1.16

Source	df	SS	MS	F	p-value
Treatments	2	121.24	60.62	52.84	0.000
Error	30	34.42	1.15		
Total	32	155.66			

$\mu_1 - \mu_2 \in (-5.12, -2.85)$.

$\mu_1 - \mu_3 \in (-0.74, 1.47)$.

$\mu_2 - \mu_3 \in (3.19, 5.50)$.

There is sufficient evidence to conclude that layout 2 is slower than the other two layouts.

11.1.17

Source	df	SS	MS	F	p-value
Treatments	2	0.4836	0.2418	7.13	0.001
Error	93	3.1536	0.0339		
Total	95	3.6372			

$\mu_1 - \mu_2 \in (-0.01, 0.22)$.

$\mu_1 - \mu_3 \in (0.07, 0.29)$.

$\mu_2 - \mu_3 \in (-0.03, 0.18)$.

There is sufficient evidence to conclude that the average particle diameter is larger at the low amount of stabilizer than at the high amount of stabilizer.

11.1.18

Source	df	SS	MS	F	p-value
Treatments	2	135.15	67.58	19.44	0.000
Error	87	302.50	3.48		
Total	89	437.66			

$\mu_1 - \mu_2 \in (-1.25, 1.04)$.

$\mu_1 - \mu_3 \in (1.40, 3.69)$.

$\mu_2 - \mu_3 \in (1.50, 3.80)$.

There is sufficient evidence to conclude that method 3 is quicker than the other two methods.

11.2   Randomized Block Designs

11.2.1

Source	df	SS	MS	F	p-value
Treatments	3	10.15	3.38	3.02	0.047
Blocks	9	24.53	2.73	2.43	0.036
Error	27	30.24	1.12		
Total	39	64.92			

11.2.2

Source	df	SS	MS	F	p-value
Treatments	7	26.39	3.77	3.56	0.0036
Blocks	7	44.16	6.31	5.95	0.0000
Error	49	51.92	1.06		
Total	63	122.47			

11.2.3

Source	df	SS	MS	F	p-value
Treatments	3	58.72	19.57	0.63	0.602
Blocks	9	2,839.97	315.55	10.17	0.0000
Error	27	837.96	31.04		
Total	39	3,736.64			

11.2.4

Source	df	SS	MS	F	p-value
Treatments	4	240.03	60.01	18.59	0.0000
Blocks	14	1,527.12	109.08	33.80	0.0000
Error	56	180.74	3.228		
Total	74	1,947.89			

11.2.5   (a)

Source	df	SS	MS	F	p-value
Treatments	2	8.17	4.085	8.96	0.0031
Blocks	7	50.19	7.17	15.72	0.0000
Error	14	6.39	0.456		
Total	23	64.75			

(b)
$$\mu_1 - \mu_2 \in \left(5.93 - 4.62 - \frac{\sqrt{0.456} \times 3.70}{\sqrt{8}}, 5.93 - 4.62 + \frac{\sqrt{0.456} \times 3.70}{\sqrt{8}} \right)$$

$$= (0.43, 2.19).$$

$$\mu_1 - \mu_3 \in \left(5.93 - 4.78 - \frac{\sqrt{0.456} \times 3.70}{\sqrt{8}}, 5.93 - 4.78 + \frac{\sqrt{0.456} \times 3.70}{\sqrt{8}} \right)$$

$$= (0.27, 2.03).$$

$$\mu_2 - \mu_3 \in \left(4.62 - 4.78 - \frac{\sqrt{0.456} \times 3.70}{\sqrt{8}}, 4.62 - 4.78 + \frac{\sqrt{0.456} \times 3.70}{\sqrt{8}} \right)$$

$$= (-1.04, 0.72).$$

11.2.6   The numbers in the "Blocks" row change (except for the degrees of freedom) and the total sum of squares changes.

11.2.7   (a) $\bar{x}_{1.} = 6.0617.$
             $\bar{x}_{2.} = 7.1967.$
             $\bar{x}_{3.} = 5.7767.$

          (b) $\bar{x}_{.1} = 7.4667.$
             $\bar{x}_{.2} = 5.2667.$
             $\bar{x}_{.3} = 5.1133.$
             $\bar{x}_{.4} = 7.3300.$
             $\bar{x}_{.5} = 6.2267.$
             $\bar{x}_{.6} = 6.6667.$

          (c) $\bar{x}_{..} = 6.345.$

          (d) $SSTr = 6.7717.$

          (e) $SSBl = 15.0769.$

          (f) $\sum_{i=1}^{k} \sum_{j=1}^{b} x_{ij}^2 = 752.1929.$

          (g) $SST = 27.5304.$

          (h) $SSE = 5.6818.$

          (i)

Source	df	SS	MS	F	p-value
Treatments	2	6.7717	3.3859	5.96	0.020
Blocks	5	15.0769	3.0154	5.31	0.012
Error	10	5.6818	0.5682		
Total	17	27.5304			

          (j) $$\mu_1 - \mu_2 \in \left(6.06 - 7.20 - \frac{\sqrt{0.5682} \times 3.88}{\sqrt{6}}, 6.06 - 7.20 + \frac{\sqrt{0.5682} \times 3.88}{\sqrt{6}} \right)$$

             $$= (-2.33, 0.05).$$

$$\mu_1 - \mu_3 \in \left(6.06 - 5.78 - \frac{\sqrt{0.5682} \times 3.88}{\sqrt{6}}, 6.06 - 5.78 + \frac{\sqrt{0.5682} \times 3.88}{\sqrt{6}}\right)$$

$$= (-0.91, 1.47).$$

$$\mu_2 - \mu_3 \in \left(7.20 - 5.78 - \frac{\sqrt{0.5682} \times 3.88}{\sqrt{6}}, 7.20 - 5.78 + \frac{\sqrt{0.5682} \times 3.88}{\sqrt{6}}\right)$$

$$= (0.22, 2.61).$$

(l) The total sample size required from each factor level (number of blocks) can be estimated as

$$n \geq \frac{4\,s^2\,q^2_{\alpha,k,\nu}}{L^2} = \frac{4 \times 0.5682 \times 3.88^2}{2.0^2} = 8.6$$

so that an additional $9 - 6 = 3$ blocks can be recommended.

11.2.8

Source	df	SS	MS	F	p-value
Treatments	3	67.980	22.660	5.90	0.004
Blocks	7	187.023	26.718	6.96	0.000
Error	21	80.660	3.841		
Total	31	335.662			

$\mu_1 - \mu_2 \in (-2.01, 3.46)$.

$\mu_1 - \mu_3 \in (-5.86, -0.39)$.

$\mu_1 - \mu_4 \in (-3.95, 1.52)$.

$\mu_2 - \mu_3 \in (-6.59, -1.11)$.

$\mu_2 - \mu_4 \in (-4.68, 0.79)$.

$\mu_3 - \mu_4 \in (-0.83, 4.64)$.

The total sample size required from each factor level (number of blocks) can be estimated as

$$n \geq \frac{4\,s^2\,q^2_{\alpha,k,\nu}}{L^2} = \frac{4 \times 3.841 \times 3.95^2}{4.0^2} = 14.98$$

so that an additional $15 - 8 = 7$ blocks can be recommended.

Note: In the remainder of this section confidence intervals for the pairwise differences of the factor level means are provided with an overall confidence level of 95%.

11.2.9

Source	df	SS	MS	F	p-value
Treatments	2	17.607	8.803	2.56	0.119
Blocks	6	96.598	16.100	4.68	0.011
Error	12	41.273	3.439		
Total	20	155.478			

$\mu_1 - \mu_2 \in (-1.11, 4.17)$.

$\mu_1 - \mu_3 \in (-0.46, 4.83)$.

$\mu_2 - \mu_3 \in (-1.99, 3.30)$.

There is *not* sufficient evidence to conclude that the calciners are operating at different efficiencies.

11.2.10

Source	df	SS	MS	F	p-value
Treatments	2	133.02	66.51	19.12	0.000
Blocks	7	1,346.76	192.39	55.30	0.000
Error	14	48.70	3.48		
Total	23	1,528.49			

$\mu_1 - \mu_2 \in (-8.09, -3.21)$.

$\mu_1 - \mu_3 \in (-4.26, 0.62)$.

$\mu_2 - \mu_3 \in (1.39, 6.27)$.

There is sufficient evidence to conclude that radar system 2 is better than the other two radar systems.

11.2.11

Source	df	SS	MS	F	p-value
Treatments	3	3,231.2	1,077.1	4.66	0.011
Blocks	8	29,256.1	3,657.0	15.83	0.000
Error	24	5,545.1	231.0		
Total	35	38,032.3			

$\mu_1 - \mu_2 \in (-8.20, 31.32)$.

$\mu_1 - \mu_3 \in (-16.53, 22.99)$.

$\mu_1 - \mu_4 \in (-34.42, 5.10)$.

$\mu_2 - \mu_3 \in (-28.09, 11.43)$.

$\mu_2 - \mu_4 \in (-45.98, -6.46)$.

$\mu_3 - \mu_4 \in (-37.65, 1.87)$.

There is sufficient evidence to conclude that driver 4 is better than driver 2.

11.2.12

Source	df	SS	MS	F	p-value
Treatments	2	7.47	3.73	0.34	0.718
Blocks	9	313.50	34.83	3.15	0.018
Error	18	199.20	11.07		
Total	29	520.17			

$\mu_1 - \mu_2 \in (-3.00, 4.60)$.

$\mu_1 - \mu_3 \in (-2.60, 5.00)$.

$\mu_2 - \mu_3 \in (-3.40, 4.20)$.

There is *not* sufficient evidence to conclude that there is any difference between the assembly methods.

11.2.13

Source	df	SS	MS	F	p-value
Treatments	4	8.462×10^8	2.116×10^8	66.55	0.000
Blocks	11	19.889×10^8	1.808×10^8	56.88	0.000
Error	44	1.399×10^8	3.179×10^6		
Total	59	29.750×10^8			

$\mu_1 - \mu_2 \in (4372, 8510)$.

$\mu_1 - \mu_3 \in (4781, 8919)$.

$\mu_1 - \mu_4 \in (5438, 9577)$.

$\mu_1 - \mu_5 \in (-3378, 760)$.

$\mu_2 - \mu_3 \in (-1660, 2478)$.

$\mu_2 - \mu_4 \in (-1002, 3136)$.

$\mu_2 - \mu_5 \in (-9819, -5681)$.

$\mu_3 - \mu_4 \in (-1411, 2727)$.

$\mu_3 - \mu_5 \in (-10228, -6090)$.

$\mu_4 - \mu_5 \in (-10886, -6748)$.

There is sufficient evidence to conclude that either agent 1 or agent 5 is the best agent.

11.2.14

Source	df	SS	MS	F	p-value
Treatments	3	10.637	3.546	2.01	0.123
Blocks	19	169.526	8.922	5.05	0.000
Error	57	100.641	1.766		
Total	79	280.805			

$\mu_1 - \mu_2 \in (-1.01, 1.21)$.

$\mu_1 - \mu_3 \in (-1.89, 0.34)$.

$\mu_1 - \mu_4 \in (-1.02, 1.20)$.

$\mu_2 - \mu_3 \in (-1.98, 0.24)$.

$\mu_2 - \mu_4 \in (-1.12, 1.11)$.

$\mu_3 - \mu_4 \in (-0.24, 1.98)$.

There is *not* sufficient evidence to conclude that there is any difference between the four formulations.

11.3   Supplementary Problems

11.3.1 $\mu_1 - \mu_2 \in (3.23, 11.57)$.

$\mu_1 - \mu_3 \in (4.32, 11.68)$.

$\mu_1 - \mu_4 \in (-5.85, 1.65)$.

$\mu_2 - \mu_3 \in (-3.44, 4.64)$.

$\mu_2 - \mu_4 \in (-13.60, -5.40)$.

$\mu_3 - \mu_4 \in (-13.70, -6.50)$.

11.3.2

Source	df	SS	MS	F	p-value
Treatments	3	1.9234	0.6411	22.72	0.000
Error	16	0.4515	0.0282		
Total	19	2.3749			

$\mu_1 - \mu_2 \in (-0.35, 0.26)$.

$\mu_1 - \mu_3 \in (0.38, 0.99)$.

$\mu_1 - \mu_4 \in (-0.36, 0.25)$.

$\mu_2 - \mu_3 \in (0.42, 1.03)$.

$\mu_2 - \mu_4 \in (-0.31, 0.30)$.

$\mu_3 - \mu_4 \in (-1.04, -0.43)$.

There is sufficient evidence to conclude that professor 3 assigns lower average grades.

11.3.3

Source	df	SS	MS	F	p-value
Treatments	3	5.77	1.92	0.49	0.690
Error	156	613.56	3.93		
Total	159	619.33			

$\mu_1 - \mu_2 \in (-1.27, 1.03)$.

$\mu_1 - \mu_3 \in (-0.82, 1.61)$.

$\mu_1 - \mu_4 \in (-1.16, 1.17)$.

$\mu_2 - \mu_3 \in (-0.64, 1.67)$.

$\mu_2 - \mu_4 \in (-0.97, 1.22)$.

$\mu_3 - \mu_4 \in (-1.55, 0.77)$.

There is *not* sufficient evidence to conclude that any of the cars is getting a better gas mileage than the others.

11.3.4

Source	df	SS	MS	F	p-value
Treatments	4	2,716.8	679.2	3.57	0.024
Blocks	5	4,648.2	929.6	4.89	0.004
Error	20	3,806.0	190.3		
Total	29	11,171.0			

There is not conclusive evidence that some brands are fresher than others.

11.3.5

Source	df	SS	MS	F	p-value
Treatments	4	10,381.4	2,595.3	25.70	0.000
Blocks	9	6,732.7	748.1	7.41	0.000
Error	36	3,635.8	101.0		
Total	49	20,749.9			

There is sufficient evidence to conclude that either fertilizer type 4 or type 5 provides the highest yield.

11.3.6

Source	df	SS	MS	F	p-value
Treatments	3	115.17	38.39	4.77	0.007
Blocks	11	4,972.67	452.06	56.12	0.000
Error	33	265.83	8.06		
Total	47	5,353.67			

There is sufficient evidence to conclude that clinic 3 is different from clinics 2 and 4.

Chapter 12

Simple Linear Regression and Correlation

12.1   The Simple Linear Regression Model

12.1.1   (a) $4.2 + (1.7 \times 10) = 21.2$.

(b) $3 \times 1.7 = 5.1$.

(c) $P(N(4.2 + (1.7 \times 5), 3.2^2) \geq 12) = 0.587$.

(d) $P(N(4.2 + (1.7 \times 8), 3.2^2) \leq 17) = 0.401$.

(e) $P(N(4.2 + (1.7 \times 6), 3.2^2) \geq N(4.2 + (1.7 \times 7), 3.2^2)) = 0.354$.

12.1.2   (a) $123.0 + (-2.16 \times 20) = 79.8$.

(b) $-2.16 \times 10 = -21.6$.

(c) $P(N(123.0 + (-2.16 \times 25), 4.1^2) \leq 60) = 0.014$.

(d) $P(30 \leq N(123.0 + (-2.16 \times 40), 4.1^2) \leq 40) = 0.743$.

(e) $P(N(123.0 + (-2.16 \times 30), 4.1^2) \leq N(123.0 + (-2.16 \times 27.5), 4.1^2)) = 0.824$.

12.2   Fitting the Regression Line

12.2.2 $\hat{\beta}_0 = 7.04$.

$\hat{\beta}_1 = 30.1$.

$\hat{\sigma}^2 = 6.37$.

$7.04 + (30.1 \times 0.5) = 22.4$.

12.2.3 $\hat{\beta}_0 = 39.5$.

$\hat{\beta}_1 = -2.04$.

$\hat{\sigma}^2 = 17.3$.

$39.5 + (-2.04 \times (-2.0)) = 43.6$.

12.2.4   (a) $\hat{\beta}_0 = -2,277$.

$\hat{\beta}_1 = 1.003$.

(b) $1.003 \times 1,000 = \$1,003$.

(c) $-2,277 + (1.003 \times 10,000) = \$7,753$.

(d) $\hat{\sigma}^2 = 774,211$.

(e) Extrapolation. Cannot predict accurately.

12.2.5   (a) $\hat{\beta}_0 = 36.19$.

$\hat{\beta}_1 = 0.2659$.

(b) $\hat{\sigma}^2 = 70.33$.

(c) Yes since $\hat{\beta}_1 > 0$.

(d) $36.19 + (0.2659 \times 72) = 55.33$.

12.2.6   (a) $\hat{\beta}_0 = 54.218$.

$\hat{\beta}_1 = -0.3377$.

(b) No, $\hat{\beta}_1 < 0$ suggests that aerobic fitness deteriorates with age.

$-0.3377 \times 5 = -1.6885$.

(c) $54.218 + (-0.3377 \times 50) = 37.33$.

(d) Extrapolation. Cannot predict accurately.

(e) $\hat{\sigma}^2 = 57.30$.

12.2.7   (a) $\hat{\beta}_0 = -29.59$.
$\hat{\beta}_1 = 0.07794$.

(b) $-29.59 + (0.07794 \times 2,600) = 173.1$.

(c) $0.07794 \times 100 = 7.794$.

(d) $\hat{\sigma}^2 = 286$.

12.2.8   (a) $\hat{\beta}_0 = -1.911$.
$\hat{\beta}_1 = 1.6191$.

(b) $1.6191 \times 1 = 1.6191$.
The expert is underestimating the times.
$-1.911 + (1.6191 \times 7) = 9.42$.

(c) Extrapolation. Cannot predict accurately.

(d) $\hat{\sigma}^2 = 12.56$.

12.2.9   (a) $\hat{\beta}_0 = 12.864$.
$\hat{\beta}_1 = 0.8051$.

(b) $12.864 + (0.8051 \times 69) = 68.42$.

(c) $0.8051 \times 5 = 4.03$.

(d) $\hat{\sigma}^2 = 3.98$.

12.3   Inferences on the Slope Parameter $\hat{\beta}_1$

12.3.1   (a) $(0.522 - (2.921 \times 0.142), 0.522 + (2.921 \times 0.142)) = (0.107, 0.937)$.

(b) The t-statistic is
$$\frac{0.522}{0.142} = 3.68$$
and the p-value is 0.002.

12.3.2   (a) $(56.33 - (2.086 \times 3.78), 56.33 + (2.086 \times 3.78)) = (48.44, 64.22)$.

(b) The t-statistic is
$$\frac{56.33 - 50.0}{3.78} = 1.67$$
and the p-value is 0.110.

12.3.3   (a) $s.e.(\hat{\beta}_1) = 0.08532$.

(b) $(1.003 - (2.145 \times 0.08532), 1.003 + (2.145 \times 0.08532)) = (0.820, 1.186)$.

(c) The t-statistic is
$$\frac{1.003}{0.08532} = 11.76$$
and the p-value is 0.000.

12.3.4   (a) $s.e.(\hat{\beta}_1) = 0.2383$.

(b) $(0.2659 - (1.812 \times 0.2383), 0.2659 + (1.812 \times 0.2383)) = (-0.166, 0.698)$.

(c) The t-statistic is
$$\frac{0.2659}{0.2383} = 1.12$$
and the p-value is 0.291.

(d) There is *not* sufficient evidence to conclude that on average trucks take longer to unload when the temperature is higher.

12.3.5   (a) $s.e.(\hat{\beta}_1) = 0.1282$.

(b) $(-\infty, -0.3377 + (1.734 \times 0.1282)) = (-\infty, -0.115)$.

(c) The t-statistic is
$$\frac{-0.3377}{0.1282} = -2.63$$
and the (two-sided) p-value is 0.017.

12.3.6 (a) $s.e.(\hat{\beta}_1) = 0.00437$.

(b) $(0.0779 - (3.012 \times 0.00437), 0.0779 + (3.012 \times 0.00437)) = (0.0647, 0.0911)$.

(c) The t-statistic is
$$\frac{0.0779}{0.00437} = 17.83$$
and the p-value is 0.000.

There is sufficient evidence to conclude that the house price depends upon the size of the house.

12.3.7 (a) $s.e.(\hat{\beta}_1) = 0.2829$.

(b) $(1.619 - (2.042 \times 0.2829), 1.619 + (2.042 \times 0.2829)) = (1.041, 2.197)$.

(c) If $\beta_1 = 1$ then the actual times are equal to the estimated times except for a constant difference of β_0.

The t-statistic is
$$\frac{1.619 - 1.000}{0.2829} = 2.19$$
and the p-value is 0.036.

12.3.8 (a) $s.e.(\hat{\beta}_1) = 0.06427$.

(b) $(0.8051 - (2.819 \times 0.06427), 0.8051 + (2.819 \times 0.06427)) = (0.624, 0.986)$.

(c) The t-statistic is
$$\frac{0.8051}{0.06427} = 12.53$$
and the p-value is 0.000.

There is sufficient evidence to conclude that resistance increases with temperature.

12.4   Inferences on the Regression Line

12.4.2 $(1392, 1400)$.

12.4.3 $(21.9, 23.2)$.

12.4.4 $(6754, 7755)$.

12.4.5 $(33.65, 41.02)$.

12.4.6 $(201.4, 238.2)$.

12.4.7 $(-\infty, 10.63)$.

12.4.8 $(68.07, 70.37)$.

12.5 Prediction Intervals for Future Response Values

12.5.1 $(1386, 1406)$.

12.5.2 $(19.7, 25.4)$.

12.5.3 $(5302, 9207)$.

12.5.4 $(21.01, 53.66)$.

12.5.5 $(165.7, 274.0)$.

12.5.6 $(-\infty, 15.59)$.

12.5.7 $(63.48, 74.96)$.

12.6 The Analysis of Variance Table

12.6.1

Source	df	SS	MS	F	p-value
Regression	1	40.53	40.53	2.32	0.137
Error	33	576.51	17.47		
Total	34	617.04			

$R^2 = \frac{40.53}{617.04} = 0.066.$

12.6.2

Source	df	SS	MS	F	p-value
Regression	1	120.61	120.61	6.47	0.020
Error	19	354.19	18.64		
Total	20	474.80			

$R^2 = \frac{120.61}{474.80} = 0.254.$

12.6.3

Source	df	SS	MS	F	p-value
Regression	1	870.43	870.43	889.92	0.000
Error	8	7.82	0.9781		
Total	9	878.26			

$R^2 = \frac{870.43}{878.26} = 0.991.$

12.6.4

Source	df	SS	MS	F	p-value
Regression	1	6.82×10^6	6.82×10^6	1.64	0.213
Error	23	95.77×10^6	4.16×10^6		
Total	24	102.59×10^6			

$R^2 = \frac{6.82 \times 10^6}{102.59 \times 10^6} = 0.06.$

12.6.5

Source	df	SS	MS	F	p-value
Regression	1	10.71×10^7	10.71×10^7	138.29	0.000
Error	14	1.08×10^7	774,211		
Total	15	11.79×10^7			

$R^2 = \frac{10.71 \times 10^7}{11.79 \times 10^7} = 0.908.$

12.6.6

Source	df	SS	MS	F	p-value
Regression	1	87.59	87.59	1.25	0.291
Error	10	703.33	70.33		
Total	11	790.92			

$R^2 = \frac{87.59}{790.92} = 0.111$.

12.6.7

Source	df	SS	MS	F	p-value
Regression	1	397.58	397.58	6.94	0.017
Error	18	1,031.37	57.30		
Total	19	1,428.95			

$R^2 = \frac{397.58}{1428.95} = 0.278$.

12.6.8

Source	df	SS	MS	F	p-value
Regression	1	90,907	90,907	318.05	0.000
Error	13	3,716	286		
Total	14	94,622			

$R^2 = \frac{90,907}{94,622} = 0.961$.

12.6.9

Source	df	SS	MS	F	p-value
Regression	1	411.26	411.26	32.75	0.000
Error	30	376.74	12.56		
Total	31	788.00			

$R^2 = \frac{411.26}{788.00} = 0.522$.

12.6.10

Source	df	SS	MS	F	p-value
Regression	1	624.70	624.70	156.91	0.000
Error	22	87.59	3.98		
Total	23	712.29			

$R^2 = \frac{624.70}{712.29} = 0.877$.

12.7   Residual Analysis

12.7.1   No suggestion that the fitted regression model is not appropriate.

12.7.2   No suggestion that the fitted regression model is not appropriate.

12.7.3   Possible suggestion of a slight reduction in the variability of the VO2-max values as age increases.

12.7.4   The observation with an area of 1,390 square feet appears to be an outlier. No suggestion that the fitted regression model is not appropriate.

12.7.5   The variability of the actual times increases as the estimated time increases.

12.7.6   Possible suggestion of a slight increase in the variability of the resistances at higher temperatures.

12.8 Variable Transformations

12.8.1 The model

$$y = \gamma_0 \, e^{\gamma_1 x}$$

is appropriate.

A linear regression can be performed with $\ln(y)$ as the dependent variable and with x as the input variable.

$\hat{\gamma}_0 = 9.12$.

$\hat{\gamma}_1 = 0.28$.

$\hat{\gamma}_0 \, e^{\hat{\gamma}_1 \times 2.0} = 16.0$.

12.8.2 The model

$$y = \frac{x}{\gamma_0 + \gamma_1 x}$$

is appropriate.

A linear regression can be performed with $\frac{1}{y}$ as the dependent variable and with $\frac{1}{x}$ as the input variable.

$\hat{\gamma}_0 = 1.067$.

$\hat{\gamma}_1 = 0.974$.

$\dfrac{2.0}{\hat{\gamma}_0 + (\hat{\gamma}_1 \times 2.0)} = 0.66$.

12.8.3 $\hat{\gamma}_0 = 8.81$.

$\hat{\gamma}_1 = 0.523$.

$\gamma_0 \in (6.84, 11.35)$.

$\gamma_1 \in (0.473, 0.573)$.

12.8.4 (b) $\hat{\gamma}_0 = 89.7$.

$\hat{\gamma}_1 = 4.99$.

(c) $\gamma_0 \in (68.4, 117.7)$.

$\gamma_1 \in (4.33, 5.65)$.

12.8.5 (b) $\hat{\gamma}_0 = 0.199$.

$\hat{\gamma}_1 = 0.537$.

(c) $\gamma_0 \in (0.179, 0.221)$.

$\gamma_1 \in (0.490, 0.584)$.

(d) $0.199 + \frac{0.537}{10.0} = 0.253$.

12.9 Correlation Analysis

12.9.3 The sample correlation coefficient is $r = 0.95$.

12.9.4 The sample correlation coefficient is $r = 0.33$.

12.9.5 The sample correlation coefficient is $r = -0.53$.

12.9.6 The sample correlation coefficient is $r = 0.98$.

12.9.7 The sample correlation coefficient is $r = 0.72$.

12.9.8 The sample correlation coefficient is $r = 0.94$.

12.10   Supplementary Problems

12.10.1   The t-ratio for β_0 is 6.86.

$s.e.(\hat{\beta}_1) = 0.299$.

The p-values are 0.000 and 0.256.

$R^2 = 0.046$.

Source	df	SS	MS	F	p-value
Regression	1	123.79	123.79	1.35	0.256
Error	28	2,574.61	91.95		
Total	29	2,698.40			

The fitted value is 12.05 and the residual is -23.32.

12.10.2   (a) $\hat{\beta}_0 = 95.77$.

$\hat{\beta}_1 = -0.1003$.

$\hat{\sigma}^2 = 67.41$.

(b) The sample correlation coefficient is $r = -0.69$.

(c) $(-0.1003 - (2.179 \times 0.0300), -0.1003 + (2.179 \times 0.0300)) = (-0.1657, -0.0349)$.

(d) The t-statistic is

$$\frac{-0.1003}{0.0300} = -3.34$$

and the p-value is 0.006.

There is sufficient evidence to conclude that the time taken to finish the test depends upon the SAT score.

(e) $-0.1003 \times 10 = -1.003$.

(f) $95.77 + (-0.1003 \times 550) = 40.6$.

$(35.81, 45.43)$.

$(22.09, 59.15)$.

(g) No suggestion that the fitted regression model is not appropriate.

12.10.3   (a) $\hat{\beta}_0 = 18.4$.

$\hat{\beta}_1 = 6.72$.

$\hat{\sigma}^2 = 89.3$.

(b) The sample correlation coefficient is $r = 0.85$.

(c) The t-statistic is

$$\frac{6.72}{0.27} = 24.9$$

and the p-value is 0.000.

There is sufficient evidence to conclude that the amount of scrap material depends upon the number of passes.

(d) $(6.72 - (1.960 \times 0.27), 6.72 + (1.960 \times 0.27)) = (6.19, 7.25)$.

(e) It increases by $6.72 \times 1 = 6.72$.

(f) $18.4 + (6.72 \times 7) = 65.4$.
$(62.8, 68.1)$.
$(46.6, 84.2)$.

(g) Observations $x = 2$, $y = 67.71$ and $x = 9$, $y = 48.17$ have standardized residuals with absolute values larger than three.

The linear model is reasonable. However, a curved model with a decreasing slope may be more appropriate.

12.10.4 $\hat{\beta}_0 = 2.74$.

$\hat{\beta}_1 = -0.0017$.

$\hat{\sigma}^2 = 0.0982$.

The sample correlation coefficient is $r = -0.31$.

The t-statistic is

$$\frac{-0.0017}{0.00053} = -3.20$$

and the p-value is 0.002.

There is sufficient evidence to conclude that on average lower grades are given in larger classes.

With an enrollment of 70 students the expected grade point average is
$2.74 + (-0.0017 \times 70) = 2.62$. and the prediction interval is $(2.00, 3.25)$.

Observation $x = 34$, $y = 1.63$ has a standardized residual with an absolute value larger than three.

12.10.5 $\hat{\beta}_0 = -19,440$.

$\hat{\beta}_1 = 98,543$.

$\hat{\sigma}^2 = 73.26 \times 10^6$.

The sample correlation coefficient is $r = 0.62$.

The t-statistic is

$$\frac{98{,}543}{12{,}708} = 7.75$$

and the p-value is 0.000.

With a winning proportion of 0.5 the fitted value is $-19{,}440 + (98{,}543 \times 0.5) = 29{,}831$. and the prediction interval is $(12754, 46909)$.

Chapter 13

Multiple Linear Regression and Nonlinear Regression

13.1 Introduction to Multiple Linear Regression

13.1.1   (a) $R^2 = 0.89$.

       (b)

Source	df	SS	MS	F	p-value
Regression	3	96.5	32.17	67.4	0.000
Error	26	12.4	0.477		
Total	29	108.9			

       (c) $\hat{\sigma}^2 = 0.477$

       (d)   The p-value is 0.000.

       (c) $(16.5 - (2.056 \times 2.6), 16.5 + (2.056 \times 2.6)) = (11.2, 21.8)$.

13.1.2   (a) $R^2 = 0.23$.

       (b)

Source	df	SS	MS	F	p-value
Regression	6	2.67	0.445	1.89	0.108
Error	38	8.95	0.2355		
Total	44	11.62			

       (c) $\hat{\sigma}^2 = 0.2355$

       (d)   The p-value is 0.108.

       (e) $(1.05 - (2.024 \times 0.91), 1.05 + (2.024 \times 0.91)) = (-0.79, 2.89)$.

13.1.3   (a) $(132.4 - (2.365 \times 27.6), 132.4 + (2.365 \times 27.6)) = (67.1, 197.7)$.

(b) The test statistic is $t = 4.80$ and the p-value is 0.002.

13.1.4   (a) $(0.954 - (2.201 \times 0.616), 0.954 + (2.201 \times 0.616)) = (-0.402, 2.310)$.

(b) The test statistic is $t = 1.55$ and the p-value is 0.149.

13.1.5   The test statistic for $H_0 : \beta_1 = 0$ is $t = 11.30$ and the p-value is 0.000.
The test statistic for $H_0 : \beta_2 = 0$ is $t = 5.83$ and the p-value is 0.000.
The test statistic for $H_0 : \beta_3 = 0$ is $t = 1.15$ and the p-value is 0.257.
Variable x_3 should be removed from the model.

13.1.6   The test statistic is $F = 1.56$ and the p-value is 0.233.

13.1.7   The test statistic is $F = 5.29$ and the p-value is 0.013.

13.1.8   (b) $\hat{y} = 7.280 - 0.313 - 0.1861 = 6.7809$.

13.1.9   (a) $\hat{y} = 104.9 + (12.76 \times 10) + (409.6 \times 0.3) = 355.38$.

(b) $(355.38 - (2.110 \times 17.6), 355.38 + (2.110 \times 17.6)) = (318.24, 392.52)$.

13.1.10   (a) $\hat{y} = 65.98 + (23.65 \times 1.5) + (82.04 \times 1.5) + (17.04 \times 2.0) = 258.6$.

(b) $(258.6 - (2.201 \times 2.55), 258.6 + (2.201 \times 2.55)) = (253.0, 264.2)$.

13.2 Examples of Multiple Linear Regression

13.2.1 (b) The variable "competitor's price" has a p-value of 0.216 and is not needed in the model.

The sample correlation coefficient between the competitor's price and sales is $r = -0.91$.

The sample correlation coefficient between the competitor's price and the company's price is $r = 0.88$.

(c) The sample correlation coefficient between the company's price and sales is $r = -0.96$.

Using the model
"sales" $= 107.4 - (3.67 \times$ "company's price")
the predicted sales are $107.4 - (3.67 \times 10.0) = 70.7$.

13.2.2 $\hat{\beta}_0 = 20.011$.

$\hat{\beta}_1 = -0.633$.

$\hat{\beta}_2 = -1.467$.

$\hat{\beta}_3 = 2.083$.

$\hat{\beta}_4 = -1.717$.

$\hat{\beta}_5 = 0.925$.

Keep all terms in the model.

It can be estimated that the fiber strength is maximized at $x_1 = -0.027$ and $x_2 = -0.600$.

13.2.3 (a) $\hat{\beta}_0 = -3,238.6$.

$\hat{\beta}_1 = 0.9615$.

$\hat{\beta}_2 = 0.732$.

$\hat{\beta}_3 = 2.889$.

$\hat{\beta}_4 = 389.9$.

(b) The variable "geology" has a p-value of 0.737 and is not needed in the model.

The sample correlation coefficient between the cost and geology is $r = 0.89$.

(c) The variable "rig-index" can also be removed from the model.

A final model
"cost" $= -3011 + (1.04 \times$ "depth") $+ (2.67 \times$ "downtime")
can be recommended.

13.2.4   A final model

"VO2-max" $= 88.8 - (0.343 \times$ "heart rate"$) - (0.195 \times$ "age"$) - (0.901 \times$ "body-fat"$)$

can be recommended.

13.3 Matrix Algebra Formulation of Multiple Linear Regression

13.3.1   (a)

$$\mathbf{Y} = \begin{pmatrix} 2 \\ -2 \\ 4 \\ -2 \\ 2 \\ -4 \\ 1 \\ 3 \\ 1 \\ -5 \end{pmatrix}$$

(b)

$$\mathbf{X} = \begin{pmatrix} 1 & 0 & 1 \\ 1 & 0 & -1 \\ 1 & 1 & 4 \\ 1 & 1 & -4 \\ 1 & -1 & 2 \\ 1 & -1 & -2 \\ 1 & 2 & 0 \\ 1 & 2 & 0 \\ 1 & -2 & 3 \\ 1 & -2 & -3 \end{pmatrix}$$

(c)

$$\mathbf{X'X} = \begin{pmatrix} 10 & 0 & 0 \\ 0 & 20 & 0 \\ 0 & 0 & 60 \end{pmatrix}$$

(d)

$$(\mathbf{X'X})^{-1} = \begin{pmatrix} 0.1000 & 0 & 0 \\ 0 & 0.0500 & 0 \\ 0 & 0 & 0.0167 \end{pmatrix}$$

(e)

$$\mathbf{X'Y} = \begin{pmatrix} 0 \\ 20 \\ 58 \end{pmatrix}$$

(g)

$$\hat{\mathbf{Y}} = \begin{pmatrix} 0.967 \\ -0.967 \\ 4.867 \\ -2.867 \\ 0.933 \\ -2.933 \\ 2.000 \\ 2.000 \\ 0.900 \\ -4.900 \end{pmatrix}$$

(h)

$$\mathbf{e} = \begin{pmatrix} 1.033 \\ -1.033 \\ -0.867 \\ 0.867 \\ 1.067 \\ -1.067 \\ -1.000 \\ 1.000 \\ 0.100 \\ -0.100 \end{pmatrix}$$

(i) $SSE = 7.933$.

(k) $s.e.(\hat{\beta}_1) = 0.238$.

$s.e.(\hat{\beta}_2) = 0.137$.

Both input variables should be kept in the model.

(l) The fitted value is

$0 + (1 \times 1) + (\frac{29}{30} \times 2) = 2.933$.

The standard error is 0.496.

$(1.76\ ,\ 4.11)$.

(m) $(0.16\ ,\ 5.71)$.

13.3.2   (a)

$$\mathbf{Y} = \begin{pmatrix} 3 \\ -5 \\ 2 \\ 4 \\ 4 \\ 6 \\ 3 \\ 15 \end{pmatrix}$$

(b)

$$\mathbf{X} = \begin{pmatrix} 1 & -3 & 0.5 \\ 1 & -2 & -3.0 \\ 1 & -1 & 0.5 \\ 1 & 0 & -1.0 \\ 1 & 0 & -1.0 \\ 1 & 1 & 1.5 \\ 1 & 2 & -1.0 \\ 1 & 3 & 3.5 \end{pmatrix}$$

(c)

$$\mathbf{X'X} = \begin{pmatrix} 8 & 0 & 0 \\ 0 & 28 & 14 \\ 0 & 14 & 27 \end{pmatrix}$$

(d)

$$(\mathbf{X'X})^{-1} = \begin{pmatrix} 0.125 & 0 & 0 \\ 0 & 0.048 & -0.025 \\ 0 & -0.025 & 0.050 \end{pmatrix}$$

(e)

$$\mathbf{X'Y} = \begin{pmatrix} 32 \\ 56 \\ 68 \end{pmatrix}$$

(f)

$$\hat{\beta} = \begin{pmatrix} 4 \\ 1 \\ 2 \end{pmatrix}$$

(g)

$$\hat{\mathbf{Y}} = \begin{pmatrix} 2 \\ -4 \\ 4 \\ 2 \\ 2 \\ 8 \\ 4 \\ 14 \end{pmatrix}$$

(h)

$$\mathbf{e} = \begin{pmatrix} 1 \\ -1 \\ -2 \\ 2 \\ 2 \\ -2 \\ -1 \\ 1 \end{pmatrix}$$

(j) $\hat{\sigma}^2 = 4$.

(k) $s.e.(\hat{\beta}_1) = 0.439$.
   $s.e.(\hat{\beta}_2) = 0.447$.
   Perhaps the variable x_1 could be dropped from the model (p-value = 0.072).

(l) The fitted value is
   $4 + (1 \times 1) + (2 \times 1) = 7$.
   The standard error is 0.832.
   (4.86 , 9.14).

(m) (1.43 , 12.57).

13.4 Evaluating Model Accuracy

13.4.1   (a)   Slight suggestion of greater variability in yields at higher temperatures.

        (b)   No unusually large standardized residuals.

        (c)   The points (90,85) and (200,702) have leverage values $h_{ii} = 0.547$.

13.4.2   (a)   The residual plots do not indicate any problems.

        (b)   If it is beneficial to add the variable "geology" to the model then there would be some pattern in this residual plot.

        (d)   The observation with a cost of 8,089.5 has a standardized residual of 2.01.

13.4.3   (a)   The residual plots do not indicate any problems.

        (b)   If it is beneficial to add the variable "weight" to the model then there would be some pattern in this residual plot.

        (d)   The observation with VO2-max = 23 has a standardized residual of -2.15.

13.6 Supplementary Problems

13.6.1   (b)

Source	df	SS	MS	F	p-value
Regression	2	2,224.8	1,112.4	228.26	0.000
Error	8	39.0	4.875		
Total	10	2,263.7			

(c) The test statistic is $t = 5.85$ and the p-value is 0.000.

(d) The fitted value is
$$18.18 - (44.90 \times 1) + (44.08 \times 1^2) = 17.36.$$
$(15.04, 19.68)$.

13.6.2   For C1 the t-ratio is -0.91 and the p-value is 0.371.

For C3 the t-ratio is -0.23 and the p-value is 0.817.

Source	df	SS	MS	F	p-value
Regression	4	6,018.0	1,504.5	31.56	0.000
Error	25	1,191.9	47.7		
Total	29	7,209.9			

The residual is -11.90.

$(-7.21, 2.87)$.

$(-17.26, 12.92)$.

13.6.3   (a)

$$\mathbf{Y} = \begin{pmatrix} 24 \\ 8 \\ 14 \\ 6 \\ 0 \\ 2 \\ -8 \\ -8 \\ -12 \\ -16 \end{pmatrix}$$

$$\mathbf{X} = \begin{pmatrix} 1 & -4 & 5 \\ 1 & -4 & -5 \\ 1 & -2 & 2 \\ 1 & -2 & -2 \\ 1 & 1 & 0 \\ 1 & 1 & 0 \\ 1 & 4 & 2 \\ 1 & 4 & -2 \\ 1 & 6 & 5 \\ 1 & 6 & -5 \end{pmatrix}$$

(b)

$$\mathbf{X}'\mathbf{X} = \begin{pmatrix} 10 & 10 & 0 \\ 10 & 146 & 0 \\ 0 & 0 & 116 \end{pmatrix}$$

(e)

$$\hat{\mathbf{Y}} = \begin{pmatrix} 21 \\ 11 \\ 12 \\ 8 \\ 1 \\ 1 \\ -6 \\ -10 \\ -9 \\ -19 \end{pmatrix}$$

$$\mathbf{e} = \begin{pmatrix} 3 \\ -3 \\ 2 \\ -2 \\ -1 \\ 1 \\ -2 \\ 2 \\ -3 \\ 3 \end{pmatrix}$$

(f) $SSE = 54$.

(h) $s.e.(\hat{\beta}_1) = 0.238$.

   $s.e.(\hat{\beta}_2) = 0.258$.

   Both input variables should be kept in the model.

(i) The fitted value is

$$4 - (3 \times 2) + (1 \times (-2)) = -4.$$

The standard error is 1.046.

(j) $(-11.02, 3.02)$.

Chapter 14

Multifactor Experimental Design and Analysis

14.1 Experiments with Two Factors

14.1.1

Source	df	SS	MS	F	p-value
fuel	1	96.33	96.33	3.97	0.081
car	1	75.00	75.00	3.09	0.117
fuel*car	1	341.33	341.33	14.08	0.006
Error	8	194.00	24.25		
Total	11	706.66			

14.1.2 (a)

Source	df	SS	MS	F	p-value
type	3	160.61	53.54	9.63	0.002
temp	2	580.52	290.26	52.22	0.000
type*temp	6	58.01	9.67	1.74	0.195
Error	12	66.71	5.56		
Total	23	865.85			

(c) With a confidence level 95%:

$\alpha_1 - \alpha_2 \in (0.26, 8.34)$.

$\alpha_1 - \alpha_3 \in (-2.96, 5.12)$.

$\alpha_1 - \alpha_4 \in (-6.97, 1.11)$.

$\alpha_2 - \alpha_3 \in (-7.26, 0.82)$.

$\alpha_2 - \alpha_4 \in (-11.27, -3.19)$.

$\alpha_3 - \alpha_4 \in (-8.05, 0.03)$.

(d) With a confidence level 95%:

$\beta_1 - \beta_2 \in (4.61, 10.89)$.

$\beta_1 - \beta_3 \in (8.72, 15.00)$.

$\beta_2 - \beta_3 \in (0.97, 7.25)$.

14.1.3   (a)

Source	df	SS	MS	F	p-value
tip	2	0.1242	0.0621	1.86	0.175
material	2	14.1975	7.0988	212.31	0.000
tip*material	4	0.0478	0.0120	0.36	0.837
Error	27	0.9028	0.0334		
Total	35	15.2723			

(c) Apart from one large negative residual there appears to be less variability in the measurements from the third tip.

14.1.4   (a)

Source	df	SS	MS	F	p-value
material	3	51.7	17.2	0.11	0.957
magnification	3	13,493.7	4,497.9	27.47	0.000
material*magnification	9	542.1	60.2	0.37	0.947
Error	80	13,098.8	163.7		
Total	95	27,186.3			

(c) Material type 3 has the least amount of variability.

14.1.5

Source	df	SS	MS	F	p-value
glass	2	3.134	1.567	0.32	0.732
acidity	1	18.201	18.201	3.72	0.078
glass*acidity	2	83.421	41.711	8.52	0.005
Error	12	58.740	4.895		
Total	17	163.496			

14.1.6

Source	df	SS	MS	F	p-value
A	2	230.11	115.06	11.02	0.004
B	2	7.44	3.72	0.36	0.710
A*B	4	26.89	6.72	0.64	0.645
Error	9	94.00	10.44		
Total	17	358.44			

The low level of ingredient B has the smallest amount of variability in the percentage improvements.

14.1.7

Source	df	SS	MS	F	p-value
design	2	3.896×10^3	1.948×10^3	0.46	0.685
color	1	0.120×10^3	0.120×10^3	0.03	0.882
Error	2	8.470×10^3	4.235×10^3		
Total	5	12.487×10^3			

14.2 Experiments with Three or More Factors

14.2.1 (d)

Source	df	SS	MS	F	p-value
drink	2	90.65	45.32	5.39	0.007
gender	1	6.45	6.45	0.77	0.384
age	2	23.44	11.72	1.39	0.255
drink*gender	2	17.82	8.91	1.06	0.352
drink*age	4	24.09	6.02	0.72	0.583
gender*age	2	24.64	12.32	1.47	0.238
drink*gender*age	4	27.87	6.97	0.83	0.511
Error	72	605.40	8.41		
Total	89	820.36			

14.2.2 (a)

Source	df	SS	MS	F	p-value
rice	2	527.0	263.5	1.72	0.193
fert	1	2,394.2	2,394.2	15.62	0.000
sun	1	540.0	540.0	3.52	0.069
rice*fert	2	311.6	155.8	1.02	0.372
rice*sun	2	2,076.5	1,038.3	6.78	0.003
fert*sun	1	77.5	77.5	0.51	0.481
rice*fert*sun	2	333.3	166.6	1.09	0.348
Error	36	5,516.3	153.2		
Total	47	11,776.5			

(b) Yes.

(c) No.

(d) Yes.

14.2.3 (a)

Source	df	SS	MS	F	p-value
add-A	2	324.11	162.06	8.29	0.003
add-B	2	5.18	2.59	0.13	0.877
conditions	1	199.28	199.28	10.19	0.005
add-A*add-B	4	87.36	21.84	1.12	0.379
add-A*conditions	2	31.33	15.67	0.80	0.464
add-B*conditions	2	2.87	1.44	0.07	0.930
add-A*add-B*conditions	4	21.03	5.26	0.27	0.894
Error	18	352.05	19.56		
Total	35	1,023.21			

The amount of additive B does not effect the expected value of the gas mileage although the variability of the gas mileage increases as more of additive-B is used.

14.2.4

Source	df	SS	MS	F	p-value
radar	3	40.480	13.493	5.38	0.009
aircraft	1	2.750	2.750	1.10	0.311
period	1	0.235	0.235	0.09	0.764
radar*aircraft	3	142.532	47.511	18.94	0.000
radar*period	3	8.205	2.735	1.09	0.382
aircraft*period	1	5.152	5.152	2.05	0.171
radar*aircraft*period	3	5.882	1.961	0.78	0.521
Error	16	40.127	2.508		
Total	31	245.362			

14.2.5 (d)

Source	df	SS	MS	F	p-value
machine	1	387.1	387.1	3.15	0.095
temp	1	29.5	29.5	0.24	0.631
position	1	1,271.3	1,271.3	10.35	0.005
angle	1	6,865.0	6,685.0	55.91	0.000
machine*temp	1	43.0	43.0	0.35	0.562
machine*position	1	54.9	54.9	0.45	0.513
machine*angle	1	1,013.6	1,013.6	8.25	0.011
temp*position	1	67.6	67.6	0.55	0.469
temp*angle	1	8.3	8.3	0.07	0.798
position*angle	1	61.3	61.3	0.50	0.490
machine*temp*position	1	21.0	21.0	0.17	0.685
machine*temp*angle	1	31.4	31.4	0.26	0.620
machine*position*angle	1	13.7	13.7	0.11	0.743
temp*position*angle	1	17.6	17.6	0.14	0.710
machine*temp*position*angle	1	87.5	87.5	0.71	0.411
Error	16	1,964.7	122.8		
Total	31	11,937.3			

14.2.6

Source	df	SS	MS	F	p-value
player	1	72.2	72.2	0.21	0.649
club	1	289.0	289.0	0.84	0.365
ball	1	225.0	225.0	0.65	0.423
weather	1	2,626.6	2,626.6	7.61	0.008
player*club	1	72.2	72.2	0.21	0.649
player*ball	1	169.0	169.0	0.49	0.488
player*weather	1	826.6	826.6	2.39	0.128
club*ball	1	5,700.3	5,700.3	16.51	0.000
club*weather	1	10.6	10.6	0.03	0.862
ball*weather	1	115.6	115.6	0.33	0.566
player*club*ball	1	22,500.0	22,500.0	65.17	0.000
player*club*weather	1	297.6	297.6	0.86	0.358
player*ball*weather	1	115.6	115.6	0.33	0.566
club*ball*weather	1	14.1	14.1	0.04	0.841
player*club*ball*weather	1	0.6	0.6	0.00	0.968
Error	48	16,571.0	345.2		
Total	63	49,605.8			

14.3 Supplementary Problems

14.3.1 (a)

Source	df	SS	MS	F	p-value
material	2	106.334	53.167	34.35	0.000
pressure	2	294.167	147.084	95.03	0.000
material*pressure	4	2.468	0.617	0.40	0.808
Error	27	41.788	1.548		
Total	35	444.756			

(c) With a confidence level 95%:

$\alpha_1 - \alpha_2 \in (2.61, 5.13)$.

$\alpha_1 - \alpha_3 \in (2.11, 4.64)$.

$\alpha_2 - \alpha_3 \in (-1.75, 0.77)$.

(d) With a confidence level 95%:

$\beta_1 - \beta_2 \in (-0.96, 1.56)$.

$\beta_1 - \beta_3 \in (-7.17, -4.65)$.

$\beta_2 - \beta_3 \in (-7.47, -4.95)$.

14.3.2

Source	df	SS	MS	F	p-value
location	1	34.13	34.13	2.29	0.144
coating	2	937.87	468.93	31.40	0.000
location*coating	2	43.47	21.73	1.46	0.253
Error	24	358.40	14.93		
Total	29	1,373.87			

14.3.3

Source	df	SS	MS	F	p-value
drug	3	593.19	197.73	62.03	0.000
severity	1	115.56	115.56	36.25	0.000
drug*severity	3	86.69	28.90	9.07	0.006
Error	8	25.50	3.19		
Total	15	820.94			

14.3.4 (c)

Source	df	SS	MS	F	p-value
furnace	1	570.38	570.38	27.77	0.000
layer	2	18.08	9.04	0.44	0.654
position	1	495.04	495.04	24.10	0.000
furnace*layer	2	23.25	11.63	0.57	0.582
furnace*position	1	18.38	18.38	0.89	0.363
layer*position	2	380.08	190.04	9.25	0.004
furnace*layer*position	2	84.25	42.13	2.05	0.171
Error	12	246.50	20.54		
Total	23	1,835.96			

14.3.5

Source	df	SS	MS	F	p-value
monomer	1	19.220	19.220	15.40	0.001
stab	1	13.781	13.781	11.04	0.004
cat	1	36.125	36.125	28.94	0.000
water	1	4.061	4.061	3.25	0.090
monomer*stab	1	0.000	0.000	0.00	1.000
monomer*cat	1	22.781	22.781	18.25	0.001
monomer*water	1	11.520	11.520	9.23	0.008
stab*cat	1	0.405	0.405	0.32	0.577
stab*water	1	0.011	0.011	0.01	0.926
cat*water	1	0.845	0.845	0.68	0.423
monomer*stab*cat	1	2.101	2.101	1.68	0.213
monomer*stab*water	1	2.000	2.000	1.60	0.224
monomer*cat*water	1	0.281	0.281	0.23	0.641
stab*cat*water	1	1.445	1.445	1.16	0.298
monomer*stab*cat*water	1	0.101	0.101	0.08	0.779
Error	16	19.970	1.248		
Total	31	134.649			

Chapter 15

Nonparametric Statistical Analysis

15.1 The Analysis of a Single Population

15.1.1   (c) It is not plausible.

       (d) It is not plausible.

       (e) $S(65) = 84$.
           The p-value is 0.064.

       (f) The p-value is 0.001.

       (g) $(65.0, 69.0)$.
           $(66.0, 69.5)$.

15.1.2   (c) A $N(1.1, 0.05^2)$ distribution is plausible while a $N(1.0, 0.05^2)$ distribution is not plausible.

       (d) $S(1.1) = 51$.
           The p-value is 0.049.

       (e) The p-values are 0.014 and 0.027.

       (f) $(1.102, 1.120)$.
           $(1.102, 1.120)$.
           $(1.101, 1.120)$.

15.1.3   The two-sided p-values for the null hypothesis $H_0 : \mu = 0.2$ are 0.004, 0.000 and 0.000.

       Confidence intervals for μ with a confidence level of at least 95% are

       $(0.207, 0.244)$

$(0.214, 0.244)$ and
$(0.216, 0.248)$.

15.1.4   The one-sided p-values for the null hypothesis $H_0 : \mu \geq 9.5$ are 0.288, 0.046 and 0.003.

15.1.5   (a) $S(18.0) = 14$.

(b) The exact p-value is
$$2 \times P(B(20, 0.5) \geq 14) \ = \ 0.115.$$

(c) $2 \times \Phi(-1.57) = 0.116$.

(d) $S_+(18.0) = 37$.

(e) $2 \times \Phi(-2.52) = 0.012$.

15.1.6   (a) $S(40) = 7$.

(b) The exact p-value is
$$2 \times P(B(25, 0.5) \leq 7) \ = \ 0.064.$$

(c) $2 \times \Phi(-2.00) = 0.046$.

(d) $S_+(40) = 241$.

(e) $2 \times \Phi(-2.10) = 0.036$.

15.1.7   The two-sided p-values for the null hypothesis $H_0 : \mu_A - \mu_B = 0$ are 0.296 and 0.300. Confidence intervals for $\mu_A - \mu_B$ with a confidence level of at least 95% are
$(-1.0, 16.0)$ and
$(-6.0, 17.0)$.

15.1.8   The two-sided p-values for the null hypothesis $H_0 : \mu_A - \mu_B = 0$ are 0.774 and 0.480. Confidence intervals for $\mu_A - \mu_B$ with a confidence level of at least 95% are
$(-6.0, 4.0)$ and
$(-4.0, 2.0)$.

15.1.9   The two-sided p-values for the null hypothesis $H_0 : \mu_A - \mu_B = 0$ are 0.003 and 0.002. Confidence intervals for $\mu_A - \mu_B$ with a confidence level of at least 95% are $(-13.0, -1.0)$ and $(-12.0, -3.5)$.

15.1.10   The two-sided p-values for the null hypothesis $H_0 : \mu_A - \mu_B = 0$ are 0.815 and 0.879. Confidence intervals for $\mu_A - \mu_B$ with a confidence level of at least 95% are $(-70.0, 80.0)$ and $(-65.0, 65.0)$.

15.1.11   The two-sided p-values for the null hypothesis $H_0 : \mu_A - \mu_B = 0$ are 0.541 and 0.721. Confidence intervals for $\mu_A - \mu_B$ with a confidence level of at least 95% are $(-13.6, 7.3)$ and $(-6.6, 6.3)$.

15.2   Comparing Two Populations

15.2.1   (c) The Kolmogorov-Smirnov statistic is $M = 0.2006$.

15.2.2   (c) The Kolmogorov-Smirnov statistic is $M = 0.376$.

15.2.3   The Kolmogorov-Smirnov statistic is $M = 0.40$.

15.2.4   (b) $S_A = 75.5$.

   (c) $$U_A = 75.5 - \frac{8 \times (8+1)}{2} = 39.5.$$

   (d) Since

$$U_A = 39.5 < \frac{mn}{2} = \frac{8 \times 13}{2} = 52$$

the data suggests that observations from population A tend to be smaller than observations from population B.

   (e) The p-value is 0.385.

15.2.5   (b) $S_A = 245$.

   (c) $$U_A = 245 - \frac{14 \times (14+1)}{2} = 140.$$

   (d) Since

$$U_A = 140 > \frac{mn}{2} = \frac{14 \times 12}{2} = 84$$

the data suggests that observations from population A tend to be larger than observations from population B.

   (e) The p-value is 0.004.

15.2.6   (b) $S_A = 215.5$.

   (c) $$U_A = 215.5 - \frac{15 \times (15+1)}{2} = 95.5.$$

   (d) Since

$$U_A = 95.5 < \frac{mn}{2} = \frac{15 \times 15}{2} = 112.5$$

the data suggests that observations from the standard treatment tend to be smaller than observations from the new treatment.

(e) The one-sided p-value is 0.247.

15.2.7   (c) The Kolmogorov-Smirnov statistic is $M = 0.218$.

(d) $S_A = 6{,}555.5$.

$$U_A = 6{,}555.5 - \frac{75 \times (75 + 1)}{2} = 3{,}705.5.$$

Since

$$U_A = 3{,}705.5 > \frac{mn}{2} = \frac{75 \times 82}{2} = 3{,}075.0$$

the data suggests that observations from production line A tend to be larger than observations from production line B.

The two-sided p-value is 0.027.

A 95% confidence interval for the difference in the population medians is $(0.003, 0.052)$.

15.3   Comparing Three or More Populations

15.3.1   (b) $\bar{r}_{1.} = 16.6$.
$\bar{r}_{2.} = 15.5$.
$\bar{r}_{3.} = 9.9$.

(c) $H = 3.60$.

(d) The p-value is

$$P(X > 3.60) = 0.165$$

where the random variable X has a chi-square distribution with $3 - 1 = 2$ degrees of freedom.

15.3.2   (a) $\bar{r}_{1.} = 10.4$.
$\bar{r}_{2.} = 26.1$.
$\bar{r}_{3.} = 35.4$.
$\bar{r}_{4.} = 12.5$.

(b) $H = 28.52$.

(c) The p-value is

$$P(X > 28.52) = 0.000$$

where the random variable X has a chi-square distribution with $4 - 1 = 3$ degrees of freedom.

15.3.3   (a) $\bar{r}_{1.} = 17.0$.
$\bar{r}_{2.} = 19.8$.
$\bar{r}_{3.} = 14.2$.
$H = 1.84$.
The p-value is

$$P(X > 1.84) = 0.399$$

where the random variable X has a chi-square distribution with $3 - 1 = 2$ degrees of freedom.

15.3.4 $\bar{r}_{1.} = 13.0$.
$\bar{r}_{2.} = 28.5$.
$\bar{r}_{3.} = 10.9$.
$H = 20.59$.
The p-value is

$$P(X > 20.59) = 0.000$$

where the random variable X has a chi-square distribution with $3 - 1 = 2$ degrees of freedom.

15.3.5 $\bar{r}_{1.} = 55.1$.

$\bar{r}_{2.} = 55.7$.

$\bar{r}_{3.} = 25.7$.

$H = 25.86$.

The p-value is

$$P(X > 25.86) \; = \; 0.000$$

where the random variable X has a chi-square distribution with $3 - 1 = 2$ degrees of freedom.

15.3.6 (b) $\bar{r}_{1.} = 1.50$.

$\bar{r}_{2.} = 2.83$.

$\bar{r}_{3.} = 1.67$.

(c) $S = 6.33$.

(d) The p-value is

$$P(X > 6.33) \; = \; 0.043$$

where the random variable X has a chi-square distribution with $3 - 1 = 2$ degrees of freedom.

15.3.7 (a) $\bar{r}_{1.} = 2.250$.

$\bar{r}_{2.} = 1.625$.

$\bar{r}_{3.} = 3.500$.

$\bar{r}_{4.} = 2.625$.

(b) $S = 8.85$.

(c) The p-value is

$$P(X > 8.85) \; = \; 0.032$$

where the random variable X has a chi-square distribution with $4 - 1 = 3$ degrees of freedom.

15.3.8 (a) $\bar{r}_{1.} = 2.429$.

$\bar{r}_{2.} = 2.000$.

$\bar{r}_{3.} = 1.571$.

$S = 2.57$.

The p-value is

$$P(X > 2.57) \; = \; 0.277$$

where the random variable X has a chi-square distribution with $3 - 1 = 2$ degrees of freedom.

15.3.9 $\bar{r}_{1.} = 1.125$.

$\bar{r}_{2.} = 2.875$.

$\bar{r}_{3.} = 2.000$.

$S = 12.25$.

The p-value is

$$P(X > 12.25) \; = \; 0.002$$

where the random variable X has a chi-square distribution with $3 - 1 = 2$ degrees of freedom.

15.3.10 $\bar{r}_{1.} = 2.4$.

$\bar{r}_{2.} = 1.7$.

$\bar{r}_{3.} = 1.9$.

$S = 2.60$.

The p-value is

$$P(X > 2.60) \; = \; 0.273$$

where the random variable X has a chi-square distribution with $3 - 1 = 2$ degrees of freedom.

15.3.11 $\bar{r}_{1.} = 4.42$.

$\bar{r}_{2.} = 2.50$.

$\bar{r}_{3.} = 1.79$.

$\bar{r}_{4.} = 1.71$.

$\bar{r}_{5.} = 4.58$.

$S = 37.88$.

The p-value is

$$P(X > 37.88) \; = \; 0.000$$

where the random variable X has a chi-square distribution with $5 - 1 = 4$ degrees of freedom.

15.3.12 $\bar{r}_{1.} = 2.375.$

         $\bar{r}_{2.} = 2.225.$

         $\bar{r}_{3.} = 3.100.$

         $\bar{r}_{4.} = 2.300.$

         $S = 5.89.$

The p-value is

$$P(X > 5.89) \;=\; 0.118$$

where the random variable X has a chi-square distribution with $4 - 1 = 3$ degrees of freedom.

15.4 Supplementary Problems

15.4.1   (c) The distribution is not plausible.

        (d) The distribution is not plausible.

        (e) $S(70) = 38$.
            The p-value is 0.011.

        (f) The p-value is 0.006.

        (g) Confidence intervals for μ with a confidence level of at least 95% are
            $(69.00, 70.00)$
            $(69.15, 69.85)$ and
            $(69.23, 70.01)$.

15.4.2   The one sided p-values are 0.005, 0.000 and 0.001.

        Confidence intervals for μ with a confidence level of at least 95% are

        $(30.9, 33.8)$

        $(31.3, 34.0)$ and

        $(30.2, 33.9)$.

15.4.3   The two-sided p-values for the null hypothesis $H_0 : \mu_A - \mu_B = 0$ are 0.115, 0.012 and 0.006.

        Confidence intervals for $\mu_A - \mu_B$ with a confidence level of at least 95% are

        $(-1.20, 0.10)$

        $(-1.05, -0.20)$ and

        $(-0.95, -0.19)$.

15.4.4   The two-sided p-values for the null hypothesis $H_0 : \mu_A - \mu_B = 0$ are 0.134, 0.036 and 0.020.

        Confidence intervals for $\mu_A - \mu_B$ with a confidence level of at least 95% are

        $(-1.00, 6.90)$

        $(0.20, 5.15)$ and

        $(0.49, 5.20)$.

15.4.5   (c) The Kolmogorov-Smirnov statistic is $M = 0.20$.

15.4.6　(c)　The Kolmogorov-Smirnov statistic is $M = 0.525$.

(d)　$S_A = 1,143$.

$$U_A = 1,143 - \frac{40 \times (40 + 1)}{2} = 323.$$

Since

$$U_A = 323 < \frac{mn}{2} = \frac{40 \times 40}{2} = 800$$

the data suggests that the heights under growing conditions A tend to be smaller than the heights under growing conditions B.

The p-value is 0.000.

$(-8.30, -3.50)$.

15.4.7　(b)　$S_A = 292$.

(c)　$$U_A = 292 - \frac{20 \times (20 + 1)}{2} = 82.$$

(d)　Since

$$U_A = 82 < \frac{mn}{2} = \frac{20 \times 25}{2} = 250$$

the data suggests that the observations tend to be smaller without anthraquinone than with anthraquinone.

(e)　The one-sided p-value is 0.000.

15.4.8　(b)　$\bar{r}_{1.} = 12.4$.

$\bar{r}_{2.} = 12.6$.

$\bar{r}_{3.} = 3.0$.

$\bar{r}_{4.} = 14.0$.

(c)　$H = 10.93$.

(d)　The p-value is

$$P(X > 10.93) = 0.012$$

where the random variable X has a chi-square distribution with $4 - 1 = 3$ degrees of freedom.

15.4.9　(a)　$\bar{r}_{1.} = 80.6$.

$\bar{r}_{2.} = 84.2$.

$\bar{r}_{3.} = 75.3$.

$\bar{r}_{4.} = 80.9.$

$H = 0.75.$

The p-value is

$$P(X > 0.75) \; = \; 0.861$$

where the random variable X has a chi-square distribution with $4 - 1 = 3$ degrees of freedom.

15.4.10   (b) $\bar{r}_{1.} = 3.500.$

$\bar{r}_{2.} = 2.500.$

$\bar{r}_{3.} = 1.583.$

$\bar{r}_{4.} = 4.000.$

$\bar{r}_{5.} = 3.417.$

(c) $S = 8.83.$

(d) The p-value is

$$P(X > 8.83) \; = \; 0.066$$

where the random variable X has a chi-square distribution with $5 - 1 = 4$ degrees of freedom.

15.4.11 $\bar{r}_{1.} = 1.7.$

$\bar{r}_{2.} = 1.5.$

$\bar{r}_{3.} = 3.5.$

$\bar{r}_{4.} = 4.2.$

$\bar{r}_{5.} = 4.1.$

$S = 27.36.$

The p-value is

$$P(X > 27.36) \; = \; 0.000$$

where the random variable X has a chi-square distribution with $5 - 1 = 4$ degrees of freedom.

15.4.12   (a) $\bar{r}_{1.} = 2.292.$

$\bar{r}_{2.} = 2.000.$

$\bar{r}_{3.} = 3.708.$

$\bar{r}_{4.} = 2.000.$

$S = 14.43.$

The p-value is

$$P(X > 14.43) \; = \; 0.002$$

where the random variable X has a chi-square distribution with $4 - 1 = 3$ degrees of freedom.

Chapter 16

Quality Control Methods

16.2 Statistical Process Control

16.2.1   (a)   The center line is 10.0 and the control limits are 9.7 and 10.3.

         (b)   The process is declared to be out of control at $\bar{x} = 9.5$ but not at $\bar{x} = 10.25$.

         (c) $P(9.7 \leq N(10.15, 0.2^2/4) \leq 10.3) = 0.9332$.

               The probability that an observation lies outside the control limits is therefore $1 - 0.9332 = 0.0668$.

               The average run length for detecting the change is $\frac{1}{0.0668} = 15.0$.

16.2.2   (a)   The center line is 0.650 and the control limits are 0.605 and 0.695.

         (b)   There is no evidence that the process is out of control at either $\bar{x} = 0.662$ or $\bar{x} = 0.610$.

         (c) $P(0.605 \leq N(0.630, 0.015^2) \leq 0.695) = 0.9522$.

               The probability that an observation lies outside the control limits is therefore $1 - 0.9522 = 0.0478$.

               The average run length for detecting the change is $\frac{1}{0.0478} = 20.9$.

16.2.3   (a) $P(\mu - 2\sigma \leq N(\mu, \sigma^2) \leq \mu + 2\sigma) = 0.9544$.

               The probability that an observation lies outside the control limits is therefore $1 - 0.9544 = 0.0456$.

         (b) $P(\mu - 2\sigma \leq N(\mu + \sigma, \sigma^2) \leq \mu + 2\sigma) = 0.8400$.

               The probability that an observation lies outside the control limits is therefore $1 - 0.8400 = 0.1600$.

The average run length for detecting the change is $\frac{1}{0.1600} = 6.25$.

16.2.4   The average run length is about $\frac{1}{1-0.9974} = 380$.

16.2.5   The probability that a point is above the center line and within the upper control limit is

$$P(\mu \leq N(\mu, \sigma^2) \leq \mu + 3\sigma) = 0.4987.$$

The probability that all eight points lie above the center line and within the upper control limit is therefore $0.4987^8 = 0.0038$. Similarly, the probability that all eight points lie below the center line and within the lower control limit is $0.4987^8 = 0.0038$. Consequently, the probability that all eight points lie on the same side of the center line and within the control limits is $2 \times 0.0038 = 0.0076$.

16.3 Variable Control Charts

16.3.1   (a) The $\bar{X}$-chart has a center line at 91.33 and control limits at 87.42 and 95.24.
The R-chart has a center line at 5.365 and control limits at 0 and 12.24.

  (b) No.

  (c) $\bar{x} = 92.6$.
$r = 13.1$.
The process can be declared to be out of control due to an increase in the variability.

  (d) $\bar{x} = 84.6$.
$r = 13.5$.
The process can be declared to be out of control due to an increase in the variability and a decrease in in the mean value.

  (e) $\bar{x} = 91.8$.
$r = 5.7$.
There is no evidence that the process is out of control.

  (f) $\bar{x} = 95.8$.
$r = 5.4$.
The process can be declared to be out of control due to an increase in the mean value.

16.3.2   (a) The $\bar{X}$-chart has a center line at 12.02 and control limits at 11.27 and 12.78.
The R-chart has a center line at 1.314 and control limits at 0 and 2.779.

  (b) Sample 8 lies above the upper control limits.

  (c) If sample 8 is removed then the following modified control charts can be employed.
The $\bar{X}$-chart has a center line at 11.99 and control limits at 11.28 and 12.70.
The R-chart has a center line at 1.231 and control limits at 0 and 2.602.

16.3.3   (a) The $\bar{X}$-chart has a center line at 2.993 and control limits at 2.801 and 3.186.
The R-chart has a center line at 0.2642 and control limits at 0 and 0.6029.

(b) $\bar{x} = 2.97$.

$r = 0.24$.

There is no evidence that the process is out of control.

16.4 Attribute Control Charts

16.4.1   The p-chart has a center line at 0.0500 and control limits at 0.0000 and 0.1154.

   (a) No.

   (b) $\dfrac{x}{100} \geq 0.1154 \quad \Rightarrow \quad x \geq 12.$

16.4.2   (a) Samples 8 and 22 are above the upper control limit on the p-chart.

   (b) If samples 8 and 22 are removed from the data set then a p-chart with a center line at 0.1400 and control limits at 0.0880 and 0.1920 is obtained.

   (c) $\dfrac{x}{400} \geq 0.1920 \quad \Rightarrow \quad x \geq 77.$

16.4.3   The c-chart has a center line at 12.42 and control limits at 1.85 and 22.99.

   (a) No.

   (b) At least 23.

16.4.4   (a) The c-chart has a center line at 2.727 and control limits at 0 and 7.682. Samples 16 and 17 lie above the upper control limit.

   (b) If samples 16 and 17 are removed then a c-chart with a center line at 2.150 and control limits at 0 and 6.549 is obtained.

   (c) At least 7.

16.5   Acceptance Sampling

16.5.1   (a) With $p_0 = 0.06$ there would be 3 defective items in the batch of $N = 50$ items. The producer's risk is 0.0005.

(b) With $p_1 = 0.20$ there would be 10 defective items in the batch of $N = 50$ items. The consumer's risk is 0.952.

Using a binomial approximation these probabilities are estimated to be 0.002 and 0.942.

16.5.2   (a) With $p_0 = 0.10$ there would be 2 defective items in the batch of $N = 20$ items. The producer's risk is 0.016.

(b) With $p_1 = 0.20$ there would be 4 defective items in the batch of $N = 20$ items. The consumer's risk is 0.912.

Using a binomial approximation these probabilities are estimated to be 0.028 and 0.896.

16.5.3   (a) The producer's risk is 0.000.

(b) The consumer's risk is 0.300.

16.5.4   (a) The producer's risk is 0.000.

(b) The consumer's risk is 0.991.

(c) If $c = 9$ then the producer's risk is 0.000 and the consumer's risk is 0.976.

16.5.5   The smallest value of c for which

$$P(B(30, 0.10) > c) \leq 0.05$$

is $c = 6$.

16.6 Supplementary Problems

16.6.1 (a) The center line is 1,250 and the control limits are 1,214 and 1,286.

(b) Yes. Yes.

(c) $P(1214 \leq N(1240, 12^2) \leq 1286) = 0.9848$.

The probability that an observation lies outside the control limits is therefore $1 - 0.9848 = 0.0152$.

The average run length for detecting the change is $\frac{1}{0.0152} = 66$.

16.6.2 (a) Sample 3 appears to have been out of control.

(b) If sample 3 is removed then the following modified control charts can be employed.

The $\bar{X}$-chart has a center line at 74.99 and control limits at 72.25 and 77.73.

The R-chart has a center line at 2.680 and control limits at 0 and 6.897.

(c) $\bar{x} = 74.01$.

$r = 3.4$.

There is no evidence that the process is out of control.

(d) $\bar{x} = 77.56$.

$r = 3.21$.

There is no evidence that the process is out of control.

16.6.3 (a) No.

(b) The p-chart has a center line at 0.0205 and control limits at 0 and 0.0474.

(c) $\frac{x}{250} \geq 0.0474 \quad \Rightarrow \quad x \geq 12$.

16.6.4 (a) Sample 13 lies above the center line of a c-chart. If sample 13 is removed then a c-chart with a center line at 2.333 and control limits at 0 and 6.916 is obtained.

(b) At least seven flaws.

16.6.5   The smallest value of c for which

$$P(B(50, 0.06) > c) \leq 0.025$$

is $c = 7$.

The consumer's risk is 0.007.

Chapter 17

Reliability Analysis and Life Testing

17.1 System Reliability

17.1.1 $r = 0.9985$.

17.1.2 $r = 0.9886$.

17.1.3   (a) $r \geq 0.9873$.

       (b) $r \geq 0.5271$.

       (c) $r \geq 0.95^{1/n}$.

           $r \geq 1 - 0.05^{1/n}$.

17.1.4   (a) The fourth component should be placed in parallel with the first component.

       (b) In general, the fourth component (regardless of the value of r_4) should be placed in parallel with the component with the smallest reliability.

17.1.5 $r = 0.9017$.

17.1.6 $r = 0.9507$.

17.2 Modeling Failure Rates

17.2.1   (a) 0.329.

(b) 0.487.

(c) 0.263.

17.2.2   (a) 0.368.

(b) 0.681.

(c) 0.424.

17.2.3   24.2 minutes.

17.2.4   (a) 0.449.

(b) 0.699.

17.2.5   (a) 0.214.

(b) 0.448.

(c) 37.5.

(d) 12.2.

17.2.6   (a) 0.034.

(b) 0.916.

(c) 22.8.

(d) 20.1.

17.2.7   (a) 0.142.

(b) 0.344.

(c) 3.54.

(d) $h(t) = 0.0469 \times t^2$.

(e) $\dfrac{h(5)}{h(3)} = 2.78$.

17.2.8   (a) 0.103.

(b) 0.307.

(c) 9.22.

(d) $h(t) = 1.423 \times 10^{-4} \times t^{3.5}$.

(e) $\dfrac{h(12)}{h(8)} = 4.13$.

17.3 Life Testing

17.3.1   (a) $(86.4, 223.6)$.

17.3.2   (a) $(8.18, 19.87)$.

17.3.3   (a) $(3.84, 9.93)$.

17.3.4   (a) $\hat{\mu} = 2.007$ and $\hat{\sigma} = 0.3536$.
         (b) 0.202.

17.3.5   (b) $(0.457, 0.833)$.

17.4   Supplementary Problems

17.4.1   (a) $n \geq 2$.

(b) $n \geq \dfrac{\ln(1-r)}{\ln(1-r_i)}$.

17.4.2 $r = 0.9890$.

17.4.3   (a) 0.156.

(b) 0.462.

17.4.4   (a) 0.207.

(b) 0.162.

(c) 86.4.

(d) $h(t) = 2.5 \times 10^{-5} \times t^{1.5}$.

(e) $\dfrac{h(120)}{h(100)} = 1.31$.

17.4.5   (a) $(98.85, 218.19)$.

17.4.6   (a) $\hat{\mu} = 2.549$ and $\hat{\sigma} = 0.2133$.

(b) 0.228.

17.4.7   (b) $(0.051, 0.331)$.